P9-DUX-467

IN THE FOOTSTEPS OF MR. KURTZ

IN THE FOOTSTEPS OF MR. KURTZ

LIVING ON THE BRINK OF DISASTER IN MOBUTU'S CONGO

MICHELA WRONG

HarperCollins*Publishers*

HarperCollins books may be purchased for educational, business, or sales promotional use. For information, please write: Special Markets Department, HarperCollins Publishers, Inc., 10 East 53rd Street, New York, NY 10022.

Originally published in 2000 in Great Britain by Fourth Estate Limited.

FIRST U.S. EDITION

Designed by Nancy B. Field

Library of Congress Cataloging-in-Publication Data
Wrong, Michela.
In the footsteps of Mr. Kurtz : living on the brink of disaster in Mobutu's Congo / Michela Wrong.— 1st U.S. ed.
p. cm.
Includes bibliographical references and index.
ISBN 0-06-018880-4
1. Congo (Democratic Republic)—Politics and government—1960–1997. 2. Congo (Democratic Republic)—Politics and government—1997–. 3. Mobutu Sese Seko, 1930–. 4. Kabila, Laurent-Dâsirâ. 5. Wrong, Michela, 1961–. I. Title.
DT658.25 .W76 2001
967.5103—dc21 00-052976

01 02 03 04 05 RRD 10 9 8 7 6 5 4 3 2 1

To Michael Holman,
who made sure the book got written

Contents

Acknowledgements

Writing this book, I relied on the hospitality of many. In Brussels, Sarah Lambert provided a home away from home; in Washington Patti Waldmeir welcomed me; in Paris I am indebted to Beatrice Lacoste; and in Geneva my uncle and aunt Mario and Anneke Musacchio were wonderful hosts. In Kinshasa I would have gone mad had it not been for the hospitality and friendship of Serge and Francis.

Journalists are not generally a sharing breed, so I am particularly grateful for help given by Massimo Alberizzi, Marina Rini, Victor Rousseau, Marie France Cros, David Goodhart and Philip Gourevitch. Steve Askins and Carole Collins, experts on Mobutu's financial arrangements, were kind enough to give me access to their papers. I am grateful to the *Financial Times* for funding so many of my African journeys and indulging my bookwriting ambitions. It was a great boon to be able to call on John Caveney, researcher at the *Financial Times*, for help.

My friends Iain Pears, Sarah and Juliette Towhidi were supportive throughout. My parents put up with my short temper and

Julian Harty, my computer-wise brother-in-law, kept me operational by salvaging the ruins of my laptop.

I have bounced questions and ideas incessantly off Peter Vandevelde, Arthur Malu Malu, Julie Mukendi and Professor Mabi Mulumba. Their patience and good humour are much appreciated.

I owe particular thanks to the US magazine *Transition*. At a time when no one else seemed interested in the project, editor Mike Vazquez was appreciative and enthusiastic, keeping me going when I was in danger of flagging. Edited excerpts from the prologue and chapters eight and twelve were first published in *Transition*'s pages.

Lastly, my thanks to Chris McGreal, Christian Jennings, Richard Dowden and, above all, Koert Lindyer, my cherished travelling companions throughout the years in Congo/Zaire.

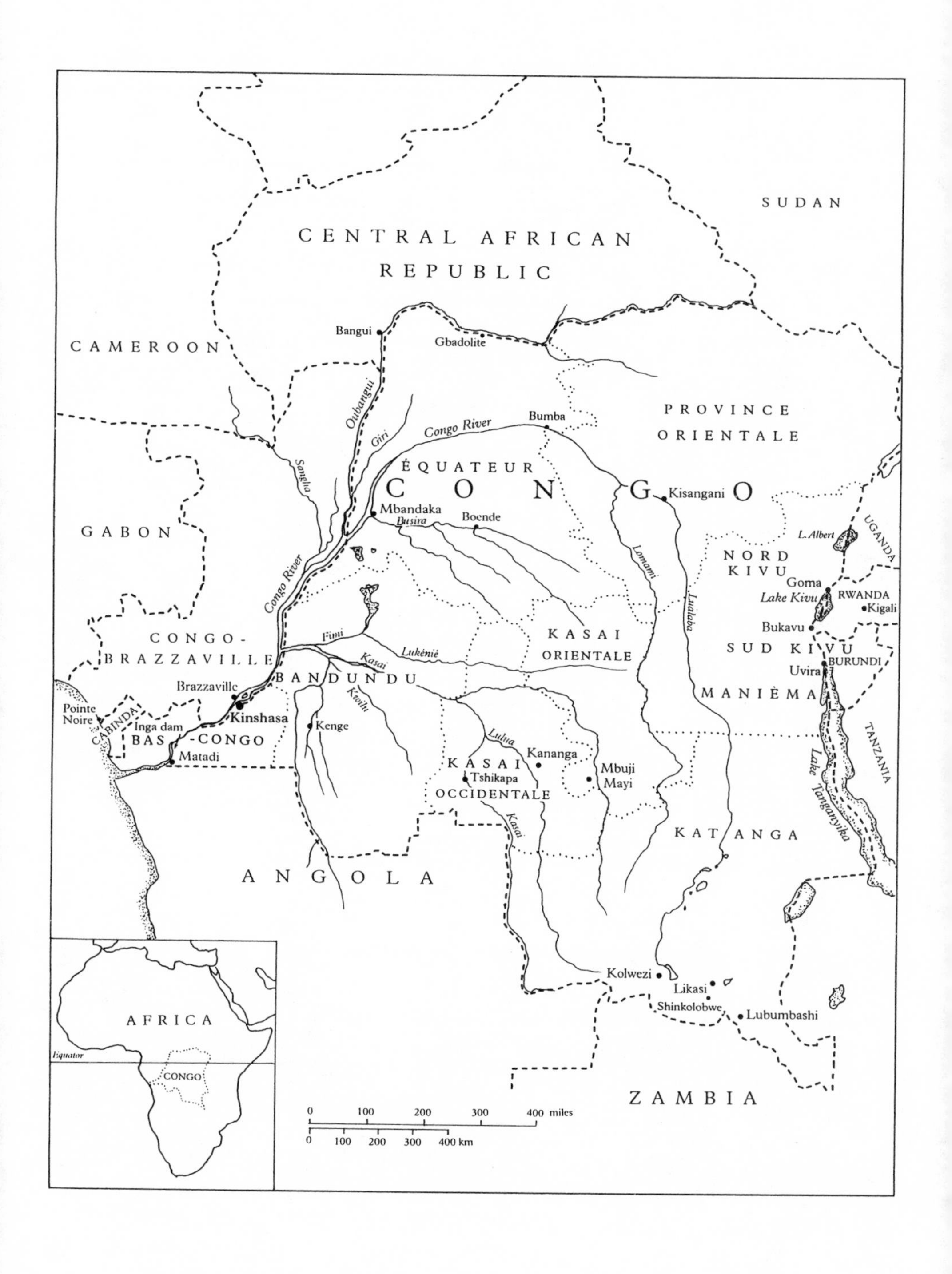

SUDAN
CENTRAL AFRICAN REPUBLIC
CAMEROON
Bangui
Gbadolite
Oubangui
Giri
Congo River
Bumba
PROVINCE ORIENTALE
Sangha
ÉQUATEUR
CONGO
Kisangani
Mbandaka
Busira
Boende
GABON
L. Albert
UGANDA
NORD KIVU
Lomami
Goma
Lake Kivu
RWANDA
Kigali
Lualaba
Congo River
Bukavu
CONGO-BRAZZAVILLE
Fimi
KASAI ORIENTALE
SUD KIVU
Lukénié
Kasai
BURUNDI
Uvira
BANDUNDU
Brazzaville
Kwilu
MANIÈMA
Pointe Noire
Kinshasa
CABINDA
Inga dam
Kenge
TANZANIA
BAS-CONGO
Lulua
Matadi
Kananga
Lake Tanganyika
KASAI
Tshikapa
Mbuji Mayi
OCCIDENTALE
Kasai
KATANGA
ANGOLA
Kolwezi
Likasi
Shinkolobwe
Lubumbashi
AFRICA
Equator
CONGO
ZAMBIA
0 100 200 300 400 miles
0 100 200 300 400 km

INTRODUCTION

'He won't be forgotten. Whatever he was, he was not common. He had the power to charm or frighten rudimentary souls into an aggravated witch-dance in his honour; he could also fill the small souls of the pilgrims with bitter misgivings: he had one devoted friend at least, and he had conquered one soul in the world that was neither rudimentary nor tainted with self-seeking. No; I can't forget him.'

Heart of Darkness
—Joseph Conrad

The feeling struck home within seconds of disembarking.

When the motor-launch deposited me in the cacophony of the quayside, engine churning mats of water hyacinth as it turned to head back across the brown expanse of oily water that was the River Zaire, I was hit by the sensation that so unnerves first-time visitors to Africa. It is that revelatory moment when white, middle-class Westerners finally understand what the rest of humanity has always known—that there are places in this world where the safety net they have spent so much of their lives erecting is suddenly whipped away, where the right accent, education, health insurance and a foreign passport—all the trappings that spell 'It Can't Happen to Me'—no longer apply, and their well-being depends on the condescension of strangers.

The pulse of apprehension drummed as I stuffed my clothes back into the ageing suitcase that had chosen the river crossing between Brazzaville and Kinshasa as the moment to split at the seams, transforming me into a truly African traveller. It quickened as a sweating young British diplomat signally failed to talk our way through the red tape and a chain of hostile policemen picked through the intimacies of my luggage, deciding which bits to keep. It subsided as we emerged from our three-hour ordeal, a little the lighter, finally crossing the magic line separating the customs area from the city.

But in truth, the quiet thud of fear would be there throughout my time in Zaire, whether I was drinking a cold Primus beer in the bustling Cité or taking tea in the green calm of a notable's patio. This ominous awareness of a world of infinite, sinister possibilities had become one of the dominant characteristics of the nation led by the man who started life as plain Joseph Désiré Mobutu, cook's son, but reinvented himself as Mobutu Sese Seko Kuku Ngbendu Wa Za Banga, 'the all-powerful warrior who goes from conquest to conquest, leaving fire in his wake'.

By the mid-1990s, Mobutu had become more noticeable by his absence than his presence, a tall, gravel-voiced figure glimpsed occasionally at official ceremonies and airport walkabouts in Kinshasa, or fielding hostile questions at a rare press conference in France with a sardonic politeness that hinted at huge world-weariness. Rattled by the army riots that had twice devastated his cities, belatedly registering the extent to which he was hated, he had withdrawn from a resentful capital to the safety of Gbadolite, his palace in the depths of the equatorial forest, to nurse his paranoia.

His impassive portrait, decked in comic-opera uniform, kept watch on his behalf, glowering from banks, shops and reception halls. 'Big Man' rule had been encapsulated in one timeless brand: leopardskin toque, Buddy Holly glasses and the carved cane so imbued with presidential force mere mortals, it was said, could never hope to lift it. He liked to be known as the Leopard, and the face of a roaring big cat was printed on banknotes, ashtrays and official letterheads. But to a population that had once hailed him as 'Papa', he was now known as 'the dinosaur', a tribute to how sclerotic his regime had become. Certainly, on a continent of dinosaur leaders, of Biya and Bongo, Mugabe and Moi, he rated as a Tyrannosaurus Rex of the breed, setting an example not to be followed. No other African autocrat had proved such a wily survivor. No other president had been presented with a country of such potential, yet achieved so little. No other leader had plundered his economy so effectively or lived the high life to such excess.

Preyed on by young men with Kalashnikovs, its administration corroded by corruption, a nation the size of Western Europe had fallen off the map of acceptable destinations. My battered copy of the Belgian Guide Nagel, picked up in a Paris bookshop, described

Kinshasa as a modern capital 'boasting all the usual attributes of Europe's great cities' and encouraged the tourist to explore its museums, monuments and 'indigenous quarters'. But that had been in 1959, when the world was a white man's oyster. Kinshasa was now a stop bypassed even by hardened travellers, where airlines avoided leaving their planes overnight for fear of what the darkness would bring. A hardship posting for diplomats, boycotted by the World Bank and IMF, it was a country every resident seemed determined to abandon, if only they could lay their hands on the necessary visa.

I would be there for the end, and for the beginning of the end.

Less than three years after my arrival, the tables were turned and I was the one to experience the curious intimacy the looter shares with his victim, rifling through Mobutu's wardrobes, touring his bathroom and making rude remarks about his taste in furniture ('African dictator' kitsch of the worst kind). Somewhere at the back of one of my drawers, there is a stolen fishknife that was once part of the presidential dining set. My companions in crime were more ambitious—they took monogrammed pillow cases, bottles of fine French wine, even a presidential oil portrait. But looters were being shot on the streets the day we paid our unannounced visit on Marshal Mobutu's villa in Goma, and I wasn't going to risk execution for a souvenir.

It was November 1996 and the new rebel movement that had suddenly risen from nowhere in the far east of Zaire had seized control of the area bordering Rwanda. For weeks the frontier crossings leading into this breathtakingly beautiful region of brooding volcanoes and misty green valleys, all rolling down to the blue waters of Lake Kivu, had been closed while the fighting went on. Then suddenly the victorious rebels opened the frontier, and a small flood of journalists who had been kicking their heels on the other side poured across.

When tour agencies were still brave enough to include Rwanda and Zaire in their African itineraries, Goma was a favourite destination for tourists visiting some of the world's last mountain gorillas. A pretty little town on the black lava foothills, it had now been torn apart by its own inhabitants, who had taken the army's exodus as the cue for some frenzied self-enrichment. Shops had been eviscerated, the main street was a mess of phone directories, glass and unused condoms, shattered toilet bowls and broken shutters. 'They've attacked me four or five times, but they just won't believe I don't have anything left to take,' gasped a ruined Lebanese trader, waiting at the border post for permission to leave. His eyes were swimming with tears.

The atmosphere was prickly. Starting what was to prove a seven-month looting and raping retreat across the country, Zairean forces had lashed out indiscriminately before pulling out, leaving corpses scattered for kilometres. No one was too sure of the identity of the rebel movement, the new bosses in town. And then there were the roaming Rwandans, whose intervention in Zaire was being denied by the government next door but was too prominent to ignore. Speaking from the corner of his mouth, a resident confirmed the outsiders' presence: 'We recognise them by their morphology.' Then he hurried away as a baby-faced Rwandan soldier—high on something and all the more sinister for the bright pink lipstick he was wearing—swaggered up to silence the blabbermouth.

Somehow, Mobutu's villa seemed the natural place to go. The road ran along the lake, snaking past walls draped in bougainvillaea, with the odd glimpse of blue water behind. We surprised a lone looter who had decided, enterprisingly, to focus on the isolated villas of the local dignitaries, rather than the overworked town centre. Thinking we were rebels, he stopped pushing a wheelbarrow on which a deep freeze was precariously balanced and ran for cover. As we drove harmlessly by, he was already returning to his task. A stolen photocopier and computer were still waiting to be taken to what, almost certainly, was a shack without electricity.

In the old days, the villa complex had been strictly off limits behind staunch metal gates manned by members of the presidential guard. Now the gates were wide open and the Zairean flag—a black fist clenching a flaming torch—lay crumpled on the ground. There had been no fight for this most symbolic of targets. No one, it was clear from the boxes of unused ammunition, the anti-tank rockets and mortar bombs carelessly stacked in the guards' quarters, had had the heart for a real showdown.

In the garage were five black Mercedes, in pristine condition, two ambulances, in case the president fell sick and a Land Rover with a podium attachment to allow him, Pope-like, to address the public. A generous allocation for a man whose visits had become increasingly rare. But like a Renaissance monarch who expected a bedroom to be provided in any of his baron's castles, Mobutu kept a dozen such mansions constantly at the ready across the country, on the off-chance of a visit that usually never came.

It was on venturing inside—could the property possibly be tripwired?—that we really began to feel like naughty children sneaking a look in their parents' bedroom, only to emerge with their illusions shattered. From outside the villa had looked the height of ostentatious luxury: all chandeliers, Ming vases, antique furniture and marble floors. Close up, almost everything proved to be fake. The vases were modern imitations, they came with price labels still attached. The Romanesque plinths were in moulded plastic, the malachite inlay painted on.

With an 'aha!' of excitement, a colleague whipped out a black and white cravat, of the type worn with the collarless 'abacost' jacket that constituted Mobutu's eccentric contribution to the world of fashion. From a distance, the cravats had always appeared complex arrangements of material, folded with meticulous care. Now I saw that they were little more than nylon bibs, held in place with tabs of Velcro. This emperor did have some clothes. But like his regime itself, they were all show and no substance.

Most poignant of all, perhaps, was the pink and burgundy suite prepared for the presidential spouse, although it was impossible to say whether this was the first lady Bobi Ladawa, or the twin sister Mobutu had, bizarrely, also taken to his bed. An outsize bottle of the perfume Je Reviens, which had probably turned rancid years ago in the African heat, stood on the mantelpiece. With their man ravaged by prostate cancer, his shambolic army collapsing like a house of cards, neither woman would ever be returning to Goma. This irreverent plundering was the only proof required of how rapidly the power established over three decades was unravelling.

Rebel uprisings, bodies rotting in the sun, a sickening megalomaniac. In newsrooms across the globe, shaking their heads over yet another unfathomable African crisis, producers and sub-editors dusted off memories of school literature courses and reached for the clichés. Zaire was Joseph Conrad's original 'Heart of Darkness', they reminded the public. How prophetic the famous cry of despair voiced by the dying Mr Kurtz at Africa's seemingly boundless capacity for bedlam and brutality had proved yet again. 'The horror, the horror.' Was nothing more promising ever to emerge from that benighted continent?

Yet when Conrad wrote *Heart of Darkness* and penned some of the most famous last words in literary history, this was very far from his intended message. The title 'Heart of Darkness' itself and the phrase 'the horror, the horror' uttered by Mr Kurtz as he expires on a steam boat chugging down the giant Congo river, probably constitute one of the great misquotations of all time.

For Conrad, the Polish seaman who was to become one of Britain's greatest novelists, *Heart of Darkness* was a book based on some very painful personal experience. In 1890 he had set out for the Congo Free State, the African colony then owned by Belgium's King Leopold II, to fill in for a steamship captain slain by tribesmen. The posting, which was originally meant to last three years but was curtailed after less than six months, was to be the most traumatic of his life. It took him nine years to digest and turn into print.

Bouts of fever and dysentery nearly killed him; his health never subsequently recovered. Always melancholic, he spent much of the time plunged into deep depression, so disgusted by his fellow whites he avoided almost all human contact. His vision of humanity was to be permanently coloured by what he found in the Congo, where declarations of philanthropy camouflaged a colonial system of unparalleled cruelty. Before the Congo, Conrad once said, 'I was a perfect animal'; afterwards, 'I see everything with such despondency—all in black'.

Mr Kurtz, whose personality haunts the book although he says almost nothing, is first presented as the best station manager of the Congo, a man of refinement and education, who can thrill crowds with his idealism and is destined for great things inside the anonymous Company 'developing' the region. Stationed 200 miles in the interior, he has now fallen sick, and a band of colleagues sets out to rescue him.

When they find him, they discover that the respected Mr Kurtz has 'gone native'. In fact, he has gone worse than native. Cut off from the Western world, inventing his own moral code and rendered almost insane by the solitude of the primeval forest, he has indulged in 'abominable satisfactions', presided 'at certain midnight dances ending with unspeakable rites' says Conrad, hinting that Kurtz has become a cannibal.

His palisade is decorated by rows of severed black heads; he has been adopted as honorary chief by a tribe whose warriors he leads on bloody village raids in search of ivory. The man who once wrote lofty reports calling for the enlightenment of the native now has a simpler recommendation: 'Exterminate all the brutes!' When he expires before the steamer reaches civilisation, corroded by fever and knowledge of his own evil, his colleagues are relieved rather than sorry—a potential embarrassment has been avoided.

Despite its slimness, the novella is one of those multilayered works whose meaning seems to shift with each new reading. By the time *Heart of Darkness* was published in 1902, the atrocities being committed by Leopold's agents in the Congo were already familiar to the public, thanks to the campaigns being waged by human rights activists of the day. So while *Heart of Darkness* is in part a psychological thriller about what makes man human, it had enough topical detail in it to carry another message to its readers. Notwithstanding the jarringly racist observations by the narrator Marlow, the way

Heart of Darkness dwells on the sense of utter alienation felt by the white man in the gloom of central Africa, the book was intended primarily as a withering attack on the hypocrisy of contemporary colonial behaviour. 'The criminality of inefficiency and pure selfishness when tackling the civilising work in Africa is a justifiable idea,' the writer told his publisher.

So when Kurtz raves against 'the horror, the horror', he is, Marlow makes clear, registering in a final lucid moment just how far he has fallen from grace. The 'darkness' of the book's title refers to the monstrous passions at the core of the human soul, lying ready to emerge when man's better instincts are suspended, rather than a continent's supposed predisposition to violence. Conrad was more preoccupied with rotten Western values, the white man's inhumanity to the black man, than, as is almost always assumed today, black savagery.

Why then, nearly a century on, has the phrase, and the title, become so misunderstood, so twisted?

The shift reflects, perhaps, the level of Western unease over Africa, a continent that has never disappointed in its capacity to disappoint: Hutu mothers killing their children by Tutsi fathers in Rwanda; the self-styled Emperor Bokassa ordering his cook to serve up his victims' bodies in Central African Republic; Liberia's rebels gleefully videotaping the torture of a former president—the terrible scenes swamp the thin trickle of good news, challenging the very notion of progress.

On a disturbing continent, no country, appropriately enough, remains more unsettling than the very birthplace of Conrad's masterpiece: the nation that was once called the Congo Free State, later metamorphosed into Zaire and has now been rebaptised the Democratic Republic of Congo.

In Mobutu's hands, the country had become a paradigm of all that was wrong with post-colonial Africa. A vacuum at the heart of the continent delineated by the national frontiers of nine neighbouring countries, it was a parody of a functioning state. Here, the anarchy and absurdity that simmered in so many other sub-Saharan nations were taken to their logical extremes. For those, like myself, curious to know what transpired when the normal rules of society were suspended, the purity appealed almost as much as it appalled. Why bother with pale imitations, diluted versions, after all, when you could drench yourself in the essence, the original?

The longer I stayed, the more fascinated I became with the man hailed as inventor of the modern kleptocracy, or government by theft. His personal fortune was said to be so immense, he could personally wipe out the country's foreign debt. He chose not to, preferring to banquet in his palaces and jet off to properties in Europe, while his citizens' average annual income had fallen below $120, leaving them dependent on their wits to survive. What could be the rationale behind such callous greed?

Zaireans had demonised him, seeing his malevolent hand behind every misfortune. From mass-murder to torture, poisoning to rape—there were few crimes not attributed to him. But if Mobutu had approached near-Satanic proportions in the popular conception, he remained the lodestar towards which every diplomat and foreign expert, opposition politician and prime ministerial candidate, turned for orientation.

Rail as it might, the population, it seemed, simply could not imagine a world without Mobutu. 'We are a peaceful people,' Zaireans would say in self-exculpation, when asked why no frenzied assailant had ever burst from the crowd during one of Mobutu's motorcades, brandishing a pistol. It was to take a foreign-backed uprising, dubbed 'an invasion' by Zaireans themselves and co-ordinated by men who did not speak the local Lingala, to rid them of the man they claimed to loathe. The passivity infuriated, eventually blurring into contempt. Every people, expatriates would shrug, deserves the leader it gets.

My attempt to understand the puzzle kept returning me to *Heart of Darkness*—not to the clichés of the headline writers, with their inverted, modernistic interpretations, but back to Conrad's original meaning.

No man is a caricature, no individual can alone bear responsibility for a nation's collapse. The disaster Zaire became, the dull political acquiescence of its people, had its roots in a history of extraordinary outside interference, as basic in motivation as it was elevated in rhetoric. The momentum behind Zaire's free-fall was generated not by one man but thousands of compliant collaborators, at home and abroad.

Exploring the Alice-in-Wonderland universe they created I would belatedly learn respect. Stumbling upon the surreal alternative systems invented by ordinary Zaireans to cope with the anarchy, exasperation would be tempered by admiration. Above all, there would be anger at what Conrad's Marlow, surveying the damage wrought by colonial conquerors who claimed to have Congo's interests at heart, described as a 'flabby, pretending, weak-eyed devil of a rapacious and pitiful folly'.

CHAPTER ONE

You can check out any time you like, but you can never leave

Kinshasa, 17 May 1997

Dear Guest,

Due to the events that have occurred last night, most of our employees have been unable to reach the hotel. Therefore, we are sorry to inform you that we will provide you only with a minimum service of room cleaning and that the laundry is only available for cleaning of your personal belongings. In advance, we thank you for your understanding and we hope that we will be able soon to assure our usual service quality.

The Management

At 3 a.m. on Saturday morning, a group of guests who had just staggered back to their rooms after a heavy drinking session in L'Atmosphère, the nightclub hidden in the bowels of Kinshasa's best hotel, heard something of a fracas taking place outside. Peering from their balconies near the top of the Tower, the modern part of the hotel where management liked to put guests paying full whack, they witnessed a scene calculated to sober them up.

Drawing up outside the Hotel Intercontinental, effectively barring all exits, were several military armoured cars, crammed with members of the Special Presidential Division (DSP), the dreaded elite unit dedicated to President Mobutu's personal protection and held responsible for the infamous Lubumbashi massacre. A black jeep with tinted windows had careered up to the side entrance and its owner—Mobutu's own son Kongulu, a DSP captain—was now levelling his sub-machine gun at the night receptionist.

Kongulu, who was later to die of AIDS, was a stocky, bearded man with a taste for fast cars, gambling and women. He left unpaid bills wherever he went with creditors too frightened to demand payment of the man who had been nicknamed 'Saddam Hussein' by Kinshasa's inhabitants. Now he was in full combat gear, bristling with grenades, two gleaming cartridge belts criss-crossed Rambo-style across his chest. And he was very, very angry.

Screaming at the receptionist, he demanded the room numbers

of an army captain and another high-ranking official staying at the Intercontinental, men he accused of betraying his father, who had fled with his family hours before rather than face humiliation at the hands of the rebel forces advancing on the capital.

Up in Camp Tsha Tshi, the barracks on the hill which housed Mobutu's deserted villa, Kongulu's fellow soldiers had already killed the only man diplomats believed was capable of negotiating a peaceful handover. With the rebels believed to be only a couple of hours' march away, Kongulu and his men were driving from one suspected hideout to another in a mood of grim fury, searching for traitors. Their days in the sun were over, they knew, but they would not go quietly. They could feel the power slipping through their fingers, but there was still time, in the moments before Mobutu's aura of invincibility finally evaporated in the warm river air, for some score-settling.

The hotel incident swiftly descended into farce, as things had a tendency to do in Zaire.

'Block the lifts,' ordered the hotel's suave Jordanian manager, determined, with a level of bravery verging on the foolhardy, to protect his guests. The night staff obediently flipped the power switch. But by the time the manager's order had got through, Kongulu and two burly soldiers were already on the sixteenth floor.

Storming from one identical door to another, unable to locate their intended victims—long since fled—and unable to descend, the death squad was reaching near-hysteria. 'Unblock the lifts, let them out, let them out,' ordered the manager, beginning to feel rattled. Incandescent with fury, the trio spilled out into the lobby. Cursing and spitting, they mustered their forces, revved their vehicles and roared off into the night, determined to slake their blood lust before dawn.

The waiting was at an end. May 17, 1997 was destined to be showdown time for Zaire. And it looked uncomfortably clear that the months of diplomatic attempts to negotiate a deal that would ease Mobutu out and rebel leader Laurent Kabila in, preventing Kinshasa from descending into a frenzy of destruction behind the departing president, had come to precisely nothing.

The fact that so many of the key episodes in what was to be Zaire's great unravelling took place in the Hotel Intercontinental was not coincidental. Africa is a continent that seems to specialise in symbolic hotels which, for months or years, are microcosms of their countries' tumultuous histories. They are buildings where atrocities are committed, *coups d'état* consecrated, embryonic rebel governments lodged, peace deals signed, and when the troubled days are over, they still miraculously come up with almond croissants, fresh coffee and CNN in most rooms.

In Rwanda, that role is fulfilled by the Mille Collines hotel, where the management stared down the Hutu militiamen bent on slaughtering terrified Tutsi guests during the 1994 genocide. In Zimbabwe, it used to be the Meikles, where armed white farmers rubbed soldiers with sanction-busters during the Smith regime. In Ethiopia it is the Hilton, where during the Mengistu years some staff doubled as government informers; in Uganda, the Nile, whose rooms once rang with the screams of suspects being tortured by Idi Amin's police.

In Congo the honour most definitely goes to the Hotel Intercontinental. I know, because I once lived there. With one room as my living quarters, another as dilapidated office and a roof-top beer crate as the perch for a satellite telex—my link with the outside world—I soon realised that the hotel, as emblematic of the regime as Mobutu's leopardskin hat, offered the perfect vantage point from which to observe the dying days of the dinosaur.

The hotel was built on a whim. On a visit to Abidjan in the Ivory Coast, President Mobutu saw the Hotel Ivoire, and decided he wanted one too. For once, his impulses were based on canny business instincts. The Intercontinental was the first five-star hotel in Kinshasa. Until the restoration of the Hotel Memling, its rival in the town centre, there was simply nowhere else to go for VIPs seeking the bland efficiency only an international hotel chain can deliver. During the prosperous 1970s, the 50 per cent government stake in the building was a share in a certified cash cow.

Constructed on a spur of land in leafy Gombe, a district of ambassadors' residences and ministries, it enjoys some of the best views in Kinshasa. To the east, the Congo river traces a lazy sweep as it emerges from Malebo pool, an expanse of water so vast that, venturing out in a small boat, you can lose sight of the opposite banks and end up wondering whether, by some miracle of geography, you have drifted out to sea.

Across the water, which is transformed into a disturbed mirror of silver and gold each sunset, gleams the distinctive concave tower that serves as the city of Brazzaville's landmark. The river, that concourse Marlow described as 'an immense snake uncoiled, with its head in the sea, its body at rest curving afar over a vast country, and its tail lost in the depths of the land' is the frontier, a fact exploited by the fishermen whose delicate pirogues languidly traverse the waterway for a spot of incidental smuggling.

Nowhere else in the world do two capitals lie so close to each other, within easy shelling distance, in fact, a feature that has been of more than merely abstract interest in the past. The proximity allows each city to act as an impromptu refugee camp when things get too hot at home. From Brazzaville to Kinshasa, from Kinshasa to Brazzaville, residents ping-pong irrepressibly from one to another—sinks, toilets and mattresses on their heads, depending on which capital is judged more dangerous at any given moment.

In peacetime, the river offers release to Kinshasa's claustrophobic expatriates. Roaring upstream in their motorboats, they picnic in the shimmering heat given off by the latest sandbank deposited by the current or scud across the waves on waterskis, weaving around the drifting islands of water hyacinth. Legend has it a European ambassador was once eaten by a crocodile while swimming and freshwater snakes are said to thrive. Yet far more ominous, for swimmers, is the steady pull of the river, the relentless tug of a vast mass of water powering relentlessly to the sea.

Some of this water has travelled nearly 3,000 miles and descended more than 5,000 feet. It has traced a huge arc curving up from eastern Zambia, heading straight north across the savannah as the Lualaba, veering west into the equatorial forest and taking in the Ubangi tributary before aiming for the Atlantic. The basin it drains rims Angola, Zambia, Tanzania, Burundi, Rwanda, Uganda, Sudan, Central African Republic and Congo-Brazzaville. The catchment area straddles the equator, ensuring that some part is always in the midst of the rainy season. Hence the river's steady flow, so strong that in theory it could cover the energy needs of central Africa and beyond. In practice, the hydroelectric dam built at Inga is working at a fraction of capacity—one of Mobutu's many white elephant projects—and even domestic demand is not being met.

The local word for river is 'nzadi': a word misunderstood and mispronounced by Portuguese explorers charting the coastline in the fifteenth century. In rebaptising Belgian Congo 'Zaire' in 1971, Mobutu was acknowledging the extent to which that waterway, the most powerful in the world after the Amazon, defines his people's identity. But what should have opened up the region has instead served to isolate it. On the map, the blue ribbon sweeping across the continent looks a promising access route. But the terrible rapids lying between the upper reaches of the Lualaba and Kisangani, Kinshasa and the sea, make nonsense of the atlas.

Looking west from the hotel, you can just glimpse the brown froth from the first of the series of falls that so appalled explorer Henry Morton Stanley when he glimpsed them in 1877. Determined to settle the dispute then raging in the West over the origins of the Nile, he had trekked across the continent from Zanzibar, losing nearly half his expedition to disease, cannibal attack and exhaustion. The calm of Malebo pool, fringed by sandy islands and a long row of white cliffs, had seemed a blessing to him and his young companion,

Frank Pocock. 'The grassy table-land above the cliffs appeared as green as a lawn, and so much reminded Frank of Kentish Downs that he exclaimed enthusiastically, "I feel we are nearing home",' wrote Stanley. In his enthusiasm Pocock, the only other white man to have survived this far into the journey, proposed naming the cliffs Dover, and the stretch of open water after Stanley. The reprieve proved shortlived. Three months later, still struggling to cross the Crystal Mountains separating the pool from the sea, Pocock went over one of the rapids and was drowned.

Leopoldville, the trading station Stanley set up here in honour of Leopold II, the Belgian King who sponsored his return to the area to 'develop' the region, was originally separate from Kinshasa, a second station established further upriver and dominated by baobab groves. The baobabs have gone now and the two stations have merged to form one inchoate city, a messy urban settlement of fits and starts that always seems about to peter away into the bush, only to sprawl that little bit further afield.

In the city's infancy, the Belgian colonisers had laid out a model city of boulevards and avenues, sports grounds and parks. But with the population now nudging five million, all thought of town planning has been abandoned, the rules of drainage and gravity ignored. Nature takes its revenge during the rainy seasons, when mini Grand Canyons open up under roads and water-logged hillsides collapse, burying inhabitants in their shacks.

'It looks as though it's survived a war and is being rebuilt,' a photographer friend, a veteran of Sarajevo, remarked after her first visit to Kinshasa. But the damage has been self-inflicted, in two rounds of looting so terrible they have become historical landmarks in people's minds, so that events are labelled as being 'avant le premier pillage' or 'après le deuxième pillage', before and after the lootings. It is Congo's version of BC and AD.

As for rebuilding, the impression given by the scaffolding and myriad work sites dotted around Kinshasa is misleading. The work has never been completed, the scaffolding will probably never be removed. Like the defunct street lamps lining Nairobi's roads, the tower blocks of Freetown, the fading boardings across Africa which advertise trips to destinations no travel company today services, it recalls another era, when a continent believed its natural trajectory pointed up instead of down.

Down in the valley lies the Cité, the pullulating popular quarters. Matonge, Makala, Kintambo: districts of green-scummed waterways, street markets and rubbish piled so high the white egrets picking through it bob above the corrugated-iron roofs. In heavy rains the open drains overflow, turning roads into rivers of black mud that exhale the warm stink of sewage. On the heights, enjoying the cooler air, are districts like Mont-Fleuri, Ma Campagne and Binza, where spiked walls conceal the mansions that housed Mobutu's elite and giant lizards in garish purple and orange do jerky press-ups by limpid blue swimming pools.

When the 'mouvanciers', as those belonging to Mobutu's presidential movement were called, ventured downhill, it was usually to the Hotel Intercontinental that they headed in their Mercedes. It was a home away from home. They liked to sit in its Atrium café in their gold-rimmed sunglasses, doing shady deals with Lebanese diamond buyers, ordering cappuccinos and talking in ostentatiously loud voices over their mobile phones while armed bodyguards loitered in the background.

They were the only ones who could afford to patronise the designer-wear shops in the hotel's arcade or hire the Junoesque whores—renowned as the most expensive in Kinshasa—who swanned along the corridors. They ran up accounts and left the management to chase payment by the government for years. Kongulu owed the casino a huge amount, but who could force a president's son to pay?

It was never a place where those who opposed the regime could feel comfortable. Mobutu's portrait stared out from above the main desk, his personality seemed to invest every echoing corridor. The Popular Movement for the Revolution (MPR), the party every Zairean at one stage was obliged to join, rented a set of rooms here and on at least one embarrassing occasion for management, a handcuffed prisoner was spotted in the lifts, being taken upstairs for interrogation.

Time the placing of your international call right and you could eavesdrop on the telephone conversations of guests down the hall, being monitored by the switchboard operators. The room cleaners showed a disproportionate level of interest in guests' comings and goings. There was always a sense of being under surveillance. 'We don't hire them as such, but what can we do if the staff work as spies?' a hotel executive once acknowledged, with a philosophical shrug of the shoulders.

By the mid-1990s the Intercontinental had, like the country itself, hit hard times. Zaire had become an international pariah and few VIPs visited Kinshasa any more. With occupancy below 20 per cent, service was stultifyingly slow. The blue dye came off the floor of the swimming pool, leaving bathers with the impression they had caught some horrible foot disease. The aroma of rotting carpet—blight of humid climates—tinged the air, the salade niçoise gave you the runs and the national power company would regularly plunge the hotel into penumbra because of unpaid bills. The first time I used the lift it shuttled repeatedly between ground floor and sixth, refusing to stop. 'Yes, we heard you ringing the alarm bell,' remarked the imperturbable receptionist when I finally won my freedom. After that I used the emergency stairs.

But there were considerations weighing against the growing tattiness, which accounted for the hotel's small population of permanent residents. We were betting on the likelihood that if Kinshasa were to be engulfed in one of its periodic bouts of pillaging, the DSP would secure the hotel. They had done so twice before, in 1991 and 1993, when the mouvanciers had slept in the conference rooms, sheltered from a frenzied populace which was dismantling their factories, supermarkets and villas.

The hotel's long-term guests were a strange bunch, representative in their way of the foreign community that washes up on African shores: misfits of the First World, sometimes intent on good works but more often escaping dubious pasts, in search of a quick killing, or simply seduced by the possibilities of misbehaviour without repercussions—that old colonial delight.

There was the ageing Belgian beauty, still sporting the miniskirts of a thirteen-year-old, who relentlessly sunbathed her way through every crisis, her appetite for ultraviolet seemingly insatiable. On the pool's fringes hovered the skinny Chinese acupuncturist, whom everyone mistook for a cook because of his starched white hat. He had come to work on an aid project in Zaire which had never seen the light of day. Given the prevalence of HIV in Kinshasa, demand for acupuncture was minimal. But he had stayed on rather than return to communist China. 'Here, it is bad. But in China, I think, maybe worse,' he confessed.

On first name terms with most of the mouvanciers was the blond, big-hearted American with a southern drawl who slopped around in flip-flops and T-shirts. Just what he was doing in Kinshasa was a mystery, but he would often use a vague, collective 'we' when referring to those in power. The Zairean staff referred to him openly as 'the CIA man', although the American embassy claimed to be unaware of his existence. Somehow, one couldn't help feeling that a real CIA man would have been a bit put out at having his role so universally recognised.

There were bored foreign pilots who flew supplies into UNITA-held territory in Angola, busting UN sanctions on salaries generous enough to merit turning a few blind eyes. 'I have told my bosses, the one thing I will never do is fly arms,' said Jean-Marie, a charming Frenchman. 'They can ask me to do anything else, but not that.' I would nod sympathetically, pretending to believe him.

Jean-Marie looked great in his pilot's uniform and spent a lot of time gently chatting up aid workers around the pool. He had shown me a photograph of his girlfriend back in France, who looked stunningly attractive but was clearly half his age. A Saint-Exupéry gone astray, he would return from trips halfway across the world—not carrying arms—and rave with Gallic lyricism about the beauty of the night sky from the pilot's cockpit. When he fell out with his bosses, he moved into a house the CIA man had started renting, although he said the mysterious goings-on there made him uneasy. One day he disappeared, never to be heard of again, and with him went the several thousand dollars it emerged he had borrowed from the CIA man and his aidworker girlfriends.

And finally, of course, there was the pony-tailed piano player. Wizened and impassive, he had been playing in the Atrium café as long as anyone could remember. He had tinkled out his lugubrious version of 'As Time Goes By' as his frame became more hunched and his hair turned from black, to first salt-and-pepper, and finally to dirty white. By May 1997, it was the piano player's puzzling absence, as much as any other event, that signalled a fundamental change was looming. A seismic shift in the world as we knew it was about to take place, and the piano player, for one, did not want to be around to see it.

The rebel movement born in Kivu in late 1996, which had triggered hoots of derisive laughter when it had pledged to overturn Mobutu, had proved far more formidable than anticipated. As it had begun capturing territory, sceptical Zaireans had gone from dismissing it as a Rwandan invasion led by a discredited Maoist to welcoming it as a liberation force. Neighbouring countries with longstanding gripes against Mobutu joined the bandwagon and the Alliance of Democratic Forces for the Liberation of Congo-Zaire (AFDL) picked up momentum.

Up in Binza, the mouvanciers had gone from haughty dismissals of the rebel problem to frantic questions: why wouldn't Mobutu *DO* something? Drained by prostate cancer, the president had curled up in his lair on the hill like a sick animal. 'When you are a soldier,' he declared, 'either you surrender or you are killed. But you don't flee.'

Days dragged into weeks. Bracing for the worst, anxious Western governments quietly pulled together a force in Brazzaville whose commandos practised the cross-river trip in high-power motorlaunches and helicopters. The diplomats were busy, juggling a stream of visa requests from the mouvanciers with preparations for the evacuation of expatriates who were stubbornly refusing to heed the increasingly forceful warnings issued over the BBC World Service.

'We've built a special cement step to allow women with high heels to get into the motor launches. And I've even got peanuts and chocolate bars ready for anyone who might starve to death while we're waiting for our men,' an ambassador proudly announced. He had gone on a trial run across the river and returned somewhat breathless. 'Door to door, it took just three and a half minutes.'

The rebels kept marching. National television broadcast footage of General Nzimbi Nzale, head of the DSP, haranguing his troops for hour after hour, ordering them to defend Mobutu to the death. The camera frame was tight and one assumed, from his hoarse tones, that he was addressing an audience of thousands. But the military made the mistake of allowing a foreign television crew to attend the same event. They filmed the general from behind, revealing a couple of dozen nose-picking soldiers, vacant-eyed, barely paying attention. Could these be the same men who had drawn up a list of strategic sites to be blown up and personalities to be assassinated once the rebels reached the city, a list leaked to Kinshasa newspapers?

In the Hotel Intercontinental the shops, anticipating the looting that traditionally preceded the rebels' arrival, first slashed the prices on their designer brands and then staged 'everything must go' sales, trying to shift stock before a more dramatic type of 'liquidation totale' occurred.

But their usual customers were no longer interested. Quietly, the mouvanciers were abandoning their villas in the hills and moving down to the Hotel Intercontinental, where they spent fitful nights, armed bodyguards perched on seats outside their rooms. You would spot them in the lobby, surrounded by matched sets of Louis Vuitton luggage, before they boarded planes and headed for properties bought years before in Belgium, France, Switzerland and South Africa in preparation for just such a day. It was almost possible to squeeze out a tiny pang of sympathy for these, the most well-heeled refugees in the world.

As for the expatriates, they had been told by their embassies to keep one holdall at the ready for the eventual evacuation, so shopping was ruled out. The designer stock stayed stubbornly put, and the evening ritual amongst journalists staying at the hotel became a window-shopping tour to mentally select which bargain to snatch as the crowds surged through the plate glass.

'What you have to realise is you'll only get the chance to go for one item,' a veteran correspondent told me with deadly seriousness. 'There won't be any time for faffing around. So it's all about focus. Quick in, quick out.' I dallied for a while over a pair of yellow lace knickers with matching bra. But in the end a tan leather jacket, worth at least $1,000, I reckoned, by Kinshasa prices, won my vote.

We were not the only ones getting light-headed with anxiety. A dinner hosted by a Zairean friend who worked at one of the ministries was a jolly, noisy meal until one of the guests called for silence. Looking around the gathering of lawyers, university professors and consultants, he raised a glass of pink champagne and reminded them that this was exactly the social class targeted for elimination after Liberia's 1980 army *coup*. 'Let us drink a toast to change, and pray we are all still here in a year's time to celebrate,' he said.

Soon after, a curfew was announced, and evening outings came to an end. Defeated soldiers and deserters were trickling into Kinshasa, hijacking the first cars they stumbled upon. It was no longer safe to venture out after dark. Instead, along with a growing number of crop-haired 'security experts' brought in by the embassies, we were confined to the Intercontinental's pizzeria, where the band laughably dubbed 'Le Best' serenaded us with a muzak medley which always featured a particularly mournful cover version of 'Hotel California'. 'You can check out any time you like, but you can never leave,' they wailed.

All airlines had now cancelled their flights to Kinshasa and the ferries had been requisitioned by the government. After weeks spent wondering whether to go or stay, the decision had been taken out of our hands.

The Hotel Intercontinental manager was finding the experience as claustrophobic as the rest of us. His appearance was as natty as ever, but his face was beginning to show the strain. 'How much longer is this going to go on? I can't eat, I can't sleep. It's giving me ulcers,' he confessed over breakfast. Fearing a siege, he had stockpiled enough food, water and diesel to cater for 2,000 people for at least a fortnight. Now he chose to combat the tension the only way he knew how: by entertaining in style. Select, candle-lit dinners were staged in the wine cellars of the hotel. Surrounded by dusty vintages, nestling in the bowels of the building, for one brief night we felt sheltered from the approaching storm.

'Do you really think these Tutsi troops are going to be as effective as people say?' asked my neighbour as we savoured the *nouvelle cuisine*. 'I suspect it's all a myth. It's easy enough beating the Zairean army out in the sticks. But surely when the rebels get to Kinshasa, and the DSP have nowhere left to run, it'll be completely different?'

It was a view I heard repeatedly, but not one I shared. I had no expectations the DSP would ever do battle. What I feared was that they would go for the soft targets, like journalists. I had developed a habit of shouting in my sleep and regretted now checking into the sixteenth floor. A precise image haunted me: looking through the spyhole in my door and seeing two DSP men, guns cocked, about to break into the room and toss me out of the window. Even if I hit the main building on the trip down, there was no way I could survive the fall from that height.

Radio Trottoir, 'pavement radio', as the city's gossip network was known, was in overdrive. There were rumours of Chinese mercenaries landing in their hundreds, of Zulu troops being called in from South Africa, of goose-stepping soldiers coming in from North Korea to save Mobutu. Also circulating were leaflets telling residents who wanted change to tie white bandanas around their foreheads when the rebels arrived as a sign of support. On the main routes into town, tanks and artillery had been set up. But with each soldier convinced a rival unit was bent on treachery, they were too busy watching each other to stop the steady flow of infiltrators into Kinshasa.

Given the steady ratcheting of tension, it was no surprise that on 15 May anyone who owned a television sat glued to their set. Since mid-afternoon a message had been running across the screen, promising an important press conference. The word on Radio

Trottoir was that Mobutu had been meeting with his generals and his departure was about to be announced.

The hours ticked by and nothing happened. The message continued to unroll. Finally, after midnight, a nervous newscaster appeared. To a rapt audience he read out a bland summary of the day's events, rounded off with a piece of homely advice: viewers should watch out for the small beetles emerging after the recent seasonal rains, which packed a particularly nasty bite.

Whatever talks had taken place in Zaire's upper echelons, commonsense had not triumphed. Mobutu, who had always warned his countrymen that 'ma tête vaut cher' ('my head won't come cheap') could not let go. When he drove to the airport the following day, heading for the jungle palace where, it was said, he planned to exhume his ancestor's bodies to save them from desecration by the rebels, he stole away in silence, having taken none of the hard decisions demanded.

And so it was that six hours after the death squad's first unwelcome visit to the Intercontinental, I found myself peering over the balcony, watching as the parking lot below filled with gleaming jeeps and flashy sports cars. Kongulu and his men were back, and this time they had arrived in force.

The lifts filled with panicking women, their hair in a mess, juggling sleepy children in pyjamas, bulging holdalls and plastic bags full of documents. Not only had we been sleeping alongside the regime's fifth columnists for the last few days, it emerged, we'd been unwitting neighbours of the DSP chiefs' extended families.

I could see their menfolk patrolling nervously up and down, toting sub-machine guns and draped in cartridge belts. They were wearing their trademark sunglasses, those gold-rimmed feminine accessories which should look comic on a man but instead manage to look as sinister as the wedding dresses and blonde wigs worn by Liberia's drugged fighters. They are the modern equivalent of the wooden masks donned around night fires by warriors preparing to do battle, which turn their wearers into something utterly alien—faceless instruments of violence capable of unspeakable acts.

We had chosen the Intercontinental because of its track record of safety. But in a shifting world order, yesterday's guardians could turn into today's hostage-takers. Looming all too vividly now was the possibility that the DSP might choose to make its last stand at Mobutu's hotel. Did our nosy room cleaner and nervous taxi driver, neither of whom had made an appearance that day, know something we didn't?

I called the British head of a security company downtown. 'Just let us know if you get too concerned and we'll come and get you,' he said breezily, as though it was the easiest thing in the world. But word came that a camera crew trying to leave the hotel had been roughed up by the DSP and turned back. Summoning assistance might simply precipitate a crisis. More hopeful news came from journalists trapped at the Hotel Memling, that other *de facto* media headquarters in the centre of town. Closer to the action, they were watching retreating Zairean soldiers streaming along the boulevards, a retreat turning into a rout. As they surrendered ground, the men were removing their uniforms to emerge as harmless civilians.

Down in the lobby, a similar remarkable metamorphosis had taken place, almost without our noticing. Uniforms and weapons had all disappeared. Scores of muscular young men were lounging about in modish tracksuits, not a hint of camouflage or khaki in sight. The Hotel Intercontinental suddenly appeared to be hosting a well-attended sports convention.

The truth dawned: the Intercontinental was not going to be the stage for a new Alamo. The DSP had laid their plans in advance and were using the hotel as a way station where they could round up their families and change into civilian clothing before heading for the river. From cursing the inaction of the Western force in Brazzaville, we went to praying they would keep away. The last thing we needed now was for the DSP's exit to be blocked.

Indeed, the DSP were encountering something of a logistical problem, as the first to flee had left their boats on the wrong side of the river. And this was when the Hotel Intercontinental suddenly justified its outrageously inflated prices, making up for all the suspect salads and blue feet, the years of skittering cockroaches and terrible muzak. From his office, our hotel manager called up his Lebanese friends and explained the situation. Swiftly a small fleet of Lebanese-owned motor-launches was assembled to ferry the new-found sports enthusiasts and their families across to Congo-Brazzaville. One convoy after another headed out, the limping wounded bringing up the rear. The hotel miraculously emptied and we heaved a sigh of relief. A showdown that could have cost hundreds of lives had been averted.

Across the deserted city, the Western security experts were at their work. On one street corner, a Belgian sharp-shooter took careful aim as a colleague ushered out a group of terrified nuns. Roaring around Kinshasa in a UN jeep, another leathery veteran had set himself the task of persuading what few soldiers remained to disarm. In the patronising tones you might use with a naughty toddler, he was telling teenagers so drunk they could barely focus to drop their rocket-grenade launchers before they did themselves a mischief.

By the river's edge lay what remained of Kongulu's sports car. Abandoned when he boarded a speed boat, it had already been stripped by looters of tyres, seats and spare parts. Along the same waterfront, Prime Minister Likulia Bolongo had also made his escape, ushered by French commandos onto a helicopter. Driving past the Hotel Memling, we noticed a dozen camera tripods laid out in a surreally neat row. The Japanese journalists, it seemed, had decided that the rebels would oblige them with a historic photo opportunity by marching down Kinshasa's main boulevard. In fact they were being a little more unpredictable, fanning through the surrounding districts. We finally stumbled upon them near the sports stadium: a group of quiet, disciplined Tutsi youths allowing themselves to be appraised by a curious crowd while they rested near a shot-out BMW. Its DSP passengers had abandoned their uniforms, but the strategy had not saved them. Riddled with bullets, they lay face-down in pools of blood.

Back at the Intercontinental, the Belgian sun-worshipper was already in her bikini, catching up on missed rays. But the hotel's official liberation did not come until the following day. Leaving the breakfast table, I had gone to see whether our taxi drivers had returned to their normal spot under the trees. And suddenly, there the rebels were. In flip-flops and bare feet, most of them no more than boys, staggering under the weight of shells and pieces of equipment, the column of AFDL fighters stretched as far as the eye could see down the Avenue des Trois Z.

Housewives ran in their dressing gowns across the lawns, brandishing cartons of Kellogg's Cornflakes and Cocopops as placatory offerings. But the adult commanders kept chivvying the exhausted 'kadogos' (little ones) along, afraid they would fall asleep as soon as they stopped moving. 'You must be tired,' sympathised an onlooker. 'Yes. I've walked all the way from Kampala,' replied one boy, artlessly spilling the beans on Uganda's involvement in the rebel uprising. 'Sshhh,' remonstrated his superior.

Abandoning their coffees, the hotel guests emerged to watch. There was a smattering of excited applause as the khaki procession wove its weary way up the hill to Binza, home of the mouvanciers and the site of Camp Tsha Tshi, Mobutu's last bolt-hole. From start to finish, the capture of a city of five million people, climax of the rebel campaign, had taken less than twenty-four hours. For the first time in history, a group of African nations had banded together to rid the region of a despot. The event was hailed as the start of an African Renaissance, spearheaded by a 'new breed' of African leader.

Over the next few days, Kinshasa made the changes appropriate to its new role as capital of the rebaptised Democratic Republic of Congo. The word 'Zaire' was removed from public buildings and road signs, leopard statues were blown up and the national flag—the flaming torch of the Mobutu era—painted over with the AFDL's blue and yellow. To jog rusty memories, newspapers printed the words to 'Debout Congolais' ('Congolese Arise'), the post-independence anthem being revived by Laurent Kabila, who traced his political lineage back to Patrice Lumumba, the country's first prime minister.

With ironic inevitability, the rebel leader who had promised to retire from the fray once Mobutu was toppled declared himself president and moved his administration into the Hotel Intercontinental. One day there was a peremptory knock at the door while I was in the shower. Looking through the spy hole I finally saw my nightmare vision made flesh: two twitchy young soldiers, rifles at the ready. But it was only the AFDL, checking for weapons, not a DSP unit intent on my defenestration.

In the hotel corridors, where the shops swiftly removed their 'sale' signs and jacked their prices back up, a new generation of lobbyists milled in search of advancement. The Atrium echoed with English and Swahili, instead of French and Lingala, and in the restaurants ragged AFDL fighters replaced the sinister DSP. But they shared their predecessors' habit of never paying. The manager's face grew taut once more. He was not amused when one of the rebels caused a bit of a ruckus at breakfast one day, carelessly dropping a grenade which rolled under the selection of almond croissants and *pains au chocolat*.

In theory, the AFDL was now in charge of one of Africa's richest states, a country blessed with diamonds and gold, copper and uranium, oil and timber. In practice, it had inherited a country reverting to the Iron Age society first encountered by the Portuguese explorers of the fifteenth century. The infrastructure was shattered, the army hopelessly divided. The state boasted more than half a million civil servants, who did little but wanted compensation for months of salary arrears. Foreign debts had accrued to the tune of $14 billion; the country had disastrous relations with all international institutions of importance and, worst of all, a population cynically inured to breaking the law.

With a simplistic rigour that could only be explained by the decades its cadres had spent outside the country, the AFDL set about the task of moral spring-cleaning. There were to be no Liberia-style executions. Instead, the new government declared the independence of the central bank, the institution Mobutu had treated as his personal cash reserve, and sacked the heads of the state enterprises Mobutu had milked for revenue. They went to join the former ministers and presidential business associates awaiting trial in Kinshasa's infamous Makala jail, specially repainted for its VIP intake.

Top of the investigators' list, of course, was Mobutu himself. The rebels had started legal proceedings well before reaching Kinshasa, firing off requests for the president's assets to be frozen in a dozen European and African countries while still on the move. Claiming he had evidence that Mobutu had appropriated a staggering $14 billion, with $8 billion of that stored in Switzerland alone, incoming Justice Minister Celestin Lwangi pledged to reverse the flight of capital.

But as the vestiges of Mobutu's reign were painted over and the Hotel Intercontinental, symbol of his rule, appropriated, the departed dinosaur did manage to exact his petty revenge. One of the last actions performed by the DSP families before heading out was to rid themselves of their Mobutu mementoes, stuffing MPR T-shirts and cloth printed with the president's face down the hotel lavatories. For the first week of the new regime, the AFDL leaders had to go outside to relieve themselves. Mobutu was literally clogging up the system.

CHAPTER TWO

Plaything for a king

'In every cordial-faced aborigine whom I meet I see a promise of assistance to me in the redemption of himself from the state of unproductiveness in which he at present lives. I look upon him with much of the same regard that an agriculturist views his strong-limbed child; he is a future recruit to the ranks of soldier-labourers. The Congo basin, could I have but enough of his class, would become a vast productive garden.'

> ***The Congo and the founding of its free state***
> **—Henry Morton Stanley**

Kinshasa possesses its own version of Ozymandias. In a field bordering the river, grounds owned by the Ministry of Planning, a grey metal giant lies ignored, his face buried in the grass. The raised arm that once beckoned flagging followers on to conquer new horizons now cradles the ground in a meaningless embrace. Too big to fit inside the warehouses holding smaller statues, this is the figure of Stanley that once towered over Mount Ngaliema, a hill overlooking Kinshasa. Congo's founder was unceremoniously dumped here in the 1970s, when Mobutu told the crowds it was time the country finally shrugged off the colonial mantle.

The anger that prompted the toppling of these grandiose monuments by Zaireans who decided they preferred a capital dotted with empty plinths to one tainted by Belgium offers a hint that Mobutu should not be regarded as *sui generis*, a monster out of time and place. Yet you will find no trace or explanation of that popular fury back in Brussels, in the museum specially constructed to commemorate a truly extraordinary colonial episode.

Built at the turn of the century on the orders of King Leopold II, the only European monarch to ever personally own an African colony, the Royal Museum for Central Africa boasts one of the largest collections of Congolese artifacts in the world. But the quantity of items stored inside this elegant building in Tervuren—the Belgian equivalent of Versailles—has done nothing to prevent a strikingly simplistic vision of history from emerging.

On the day I visited, the woman handing out tickets inside the marble-lined entrance hall seemed surprised I wanted to see the permanent collection, rather than a special exhibition of West African masks on temporary display. Strolling under the gilded cupolas and tip-tapping my way through the halls designed by French architect Charles Girault, Leopold's favourite, I began to see why even its staff might regard the museum as an anachronism and feel a sense of relief that a large number of the exhibits were currently hidden from view, undergoing refurbishment.

Political correctness, the modern sense that colonialism is something to be regretted rather than gloried in, had made the barest of inroads here. King Leopold's bust, with its unmistakable spade-shaped beard and beak nose, stared with proprietary ferocity from frozen courtyard and chilly hall. Under his watchful eye, history was still being sieved through the mental filter of the nineteenth-century capitalist and driven missionary—colonialism as economic opportunity and soul-saving expedition, all wrapped up into one convenient package.

One section, dedicated to Congo's flora and fauna, displayed scraps, sheets and lumps of natural rubber. But there was no mention of the methods used to extract the raw material or ensure a steady supply back to Europe. Wall paintings showed Congo's jungle being stripped to make room for copper mines, but the struggle over mineral assets between Belgium and the post-independence government did not feature. Was it a symbolic accident or deliberate, I wondered, that the lights in the rooms displaying the battered suitcase and worn khaki bag used by Stanley were barely working, discouraging any lingering over Congo's controversial pioneer?

Sly omission blurred effortlessly into blatant wishful thinking. In the Memorial Hall, where the paint was peeling off the ceiling, labels promised to reveal 'the King's intentions towards the Congo'. But the anti-slavery medals struck at Leopold's behest made the same point as the rusting slave chains in the glass cases and the melodramatic tableaux vivants, all buxom negro wenches and noble savages wincing under the whip of the sneering Arab overseer. Leopold, it seemed, colonised the Congo not for commercial reasons or vainglorious imperialist ambition, but to snuff out the barbaric slave trade that for centuries had robbed central Africa of its strongest and its best.

I had expected rose-coloured spectacles, but this complacent rewriting of Belgium's past took me by surprise. No explanation here, then, for why things went so wrong under Mobutu. This was a tale—the wall frieze commemorating the hundreds of young Belgians who found their graves in the Congo Free State made clear—of selfless commitment and higher motives.

From this self-satisfied tableau, one item nonetheless grabbed my attention. Under the roll-call of dead heroes, an 1884 painting by Edouard Manduau, a painter unknown to me, injected an incongruous note. The artist, who had clearly been somewhat disturbed by his brush with the Congo, had painted a native being held to a post. On his knees, writhing, he is being whipped until the blood flows down his back. Looking on without expression is a white man, scientifically taking notes.

In the whole museum, it was the only object on display that had the sour ring of truth. Those bright oils, that unexpected depiction of what was clearly an everyday, a banal event, pointed in a very different direction, one that would show how the seeds of Mobutism found fertile ground in which to sprout.

Jules Marchal knew all about watching coolly as a man was whipped. As a young district commissioner working in the Congo in the 1950s, he used to order labourers who had failed to meet the cotton quotas set by the Belgian state to be punished with the chicotte, a whip made from a strip of hippopotamus hide that had been dried in the sun. Applied sparingly, it flayed the skin and left permanent scars; used enthusiastically, it could kill.

'We would tour the country, taking our prison with us and then we'd call the villagers to assemble and we would beat three or four of our prisoners to show them what could happen to them,' he recalled, with a rueful shake of the head. 'I used that punishment very sparingly. But its effect was terrible. We were so proud to be members of the administrative service, we felt so powerful. But all our power had its roots in the chicotte.'

Shame and guilt have a long reach. Nearly half a century after the events, Marchal was still trying to expunge what he did as a thoughtless young administrator flush with the excitement of an exotic posting and overwhelmed by new responsibilities. Long since retired, he had dedicated the previous twenty years to contradicting the version of history presented at the Royal Museum for Central Africa, a whitewashing so clumsy it prompted an explosion of exasperated contempt. 'It's ridiculous! They even show an Arab trader whipping a slave! Absurd,' he snorted.

I had spotted Mr Marchal's name in the historical section of one of Brussels's bookshops, something of a miracle in itself, I was subsequently to discover, given his self-imposed low profile. His name had also cropped up in *King Leopold's Ghost*, the bestseller by US author Adam Hochschild, which was creating a stir amongst the Brussels intelligentsia in 1998. After my visit to the museum, I wanted to meet the man campaigning, virtually single-handed, to awaken a slumbering national conscience.

He had given me careful instructions over the phone, speaking with that slight Belgian twang that always sounds vaguely comic to anyone used to hearing French as spoken by Parisians. 'You want to get off at St Truiden. But make sure if you take the train to Liège that you sit in the right part, as the train splits in two and some of my visitors have gone missing that way.'

An hour and a half out of the capital, I was already a world away from the smart shopping streets of French-speaking Brussels. This was fruit-producing Flanders, proud of its Flemish identity and language, resentful and suspicious of Francophone dominance. The train slid past frosty piles of mangelwurzels, snow-dusted fields and rows of denuded orchards, stopping at every sleepy station.

Now a portly pensioner, Mr Marchal had a distinguished career behind him. After nearly two decades in Zaire, he became a diplomat, rising to the rank of ambassador. His were not the easy postings: he served in Sierra Leone, Ghana, Chad, Niger and Liberia. His wife, who nonetheless remembered their years in Africa with huge nostalgia, still drove the ageing blue Mercedes that was the ambassador's car on their last foreign assignment.

His earlier responsibilities made his new role as iconoclast all the more unexpected. For Mr Marchal, the former career diplomat, was busy energetically kicking the system that had sustained him. Trawling through the national archives, basing his findings on official memoranda, private correspondence, diaries kept by Belgian colonial agents, he was bent on exposing what he believed was the most brutal colonial system ever practised on a continent which saw more than its fair share of oppressive regimes.

While he worked with passionate commitment, he felt unhappy enough about the devastating light his discoveries shed on his former employers to shun the public stage. His first books had been published under a pseudonym. Some, printed by a company set up by his wife, verged on vanity publishing. Resolutely factual, the bare bones on which other, more florid writers—Mr Marchal hoped—would some day base their work, the volumes only featured on the shelves of the largest and most specialised Belgian bookshops. In the absence of active promotion, sales of 700 counted as a good result and Marchal was happy to hand out remaindered stock. 'I have to tell these things because they are true, I want to put history right. But I cannot promote my message as an ordinary author does. It is too sad,' he explained. 'Whatever you do, please don't present me as a traitor who is trying to bring down my country.'

Marchal had been accused by academic contemporaries studying the era of drawing up a 'personalised charge sheet'. Indeed, he was near-obsessed with the qualities, or lack of them, of the man he saw as holding the key to Congo's dark story. Certainly, the huge central African land mass that today occupies 905,000 square miles, nearly eighty times the size of Belgium, its colonial master, would never have been defined as a nation at all had it not been for the determination of the Duke of Brabant to acquire a colony.

Even as a young man, waiting in the wings for his father to die, the man who was to become Leopold II had taken careful note of how England, Spain, Portugal and the Netherlands had all built their power and wealth on a panoply of colonies, using foreign resources to rise above what often seemed the limitations of geography and natural assets.

His country was young, its sense of self-identity distinctly shaky. He was only the second monarch of an independent Belgian state, whose people had staged a revolution in 1830, turning their backs on centuries of Spanish, Austrian, French and Dutch rule. Despite a distinct lack of enthusiasm on the part of the population, he was determined to use a colony to transform his tiny country, divided by religion and language, into a world power commanding respect.

'No country has had a great history without colonies,' Leopold wrote to a collaborator. 'Look at the history of Venice, of Rome and Ancient Greece. A complete country cannot exist without overseas possessions and activity.' Scouring the world, he had looked at China, Guatemala, Fiji, Sarawak, the Philippines and Mozambique as possible candidates, but had been stymied at every turn. Then, cantering to the rescue like a moustachioed crusader, had come Henry Morton Stanley.

Stanley was a poor Briton who had emigrated to America, where he had reinvented himself as a war correspondent known for his racy copy and fearlessness under fire. An illegitimate child, he had been abandoned by his mother and sent to the workhouse, circumstances that left him with a deep need to prove himself. Fated to spend his life in a swirl of controversy, Stanley had first seized the public's imagination by penetrating darkest Africa in 1871 and tracking down David Livingstone, the British missionary who had gone missing five years earlier. Their legendary meeting was one of the great journalistic scoops of all time.

In 1877 he pulled off an even more impressive feat. Proposing to settle the dispute that had festered for years between British explorers John Speke and Richard Burton over the source of the Nile, he set off once more from Zanzibar, tracing the course of the Lualaba river for 1,500 miles. Braving rapids, ambushes, smallpox and starvation, he followed the river, emerging at the Atlantic Ocean after a journey that lasted nearly three years. He had not only established that the Lualaba had no connection with the Nile, which he had shown to spring from Lake Victoria, he had also opened up a huge swathe of central Africa until then known only to the 'Arab' merchants (in actual fact Swahili-speaking, Moslem traders from Africa's east coast) to greedy Western eyes.

In the books Stanley wrote after each extraordinary trip he showed a near-obsession with the dangers posed by perspiration and sodden underwear, which he blamed for malarial chills. But his eccentricities did not prevent him from accurately sizing up the potential of the land he had passed through. Its forests were full of precious woods and ivory-bearing elephants. Its fertile soils supported palm oil, gums and, most significantly, wild rubber, about to come into huge demand with the invention of the pneumatic tyre. Its inhabitants presented a ready market for European goods and, once the rapids were passed, the river offered a huge transport network stretching across central Africa.

Stanley was far from being the first white man to reach this part of central Africa. Late fifteenth-century emissaries from Portugal, looking for the fabled black Christian empire of Prester John, had stumbled on the Kongo kingdom, a Bantu empire spreading across what is today northern Angola, western Congo and edging into Congo-Brazzaville.

A feudal society led by the ManiKongo, this kingdom proved surprisingly open to the arrival of the white man, perhaps encouraged by a spiritual system which identified white, the skin colour of these strange visitors, as sacred. It had welcomed missionaries, embraced Christianity and entered into alliance with the Portuguese. But by the time Stanley was tracing the course of the river, the Kongo kingdom had been in decline for more than two centuries, devastated by endless wars of succession, attacks by hostile tribes and, above all, the flourishing slave trade.

Although it was clearly in his interest to play up the horrors of what he found, for it made the alternative of colonial subjugation seem so much more attractive, Stanley appears to have been genuinely horrified at the damage the 'Arabs' had wrought along the river.

'The slave traders admit that they have only 2300 captives in their fold, yet they have raided through the length and breadth of a country larger than Ireland, bearing fire and spreading carnage with lead and iron,' he reported in *The Congo and the founding of its free state*. 'Both banks of the river show that 118 villages, and forty-three districts have been devastated, out of which is only educed this scant profit of 2300 females and children and about 2000 tusks of ivory . . . The outcome from the territory with its million of souls is 5000 slaves, obtained at the cruel expense of 33000 lives!'

But his hopes that Britain, his mother country, would seize the opportunities presented were dashed. With London refusing to take the bait, King Leopold II stepped in. One of the last pieces of unclaimed land in a continent being portioned off by France, Portugal, Britain and Germany, Congo fitted his requirements perfectly. Leopold recruited Stanley to return to the Congo, set up a base there and establish a chain of trading stations along the navigable main stretch of the river which would allow the European sovereign to claim the region's riches.

Stanley found himself in a race against Count Pierre Savorgnan de Brazza, a naval officer who was energetically signing up local chiefs on France's behalf. With the northern shoreline lost to him—hence the eventual establishment of French Congo, with Brazzaville as its capital—Stanley had to content himself with the southern shore of the river, pushing his treaties on hundreds of chieftains. Leopold's insignia—the gold star on a blue background later, bizarrely, revived by the anti-colonial Laurent Kabila—was raised over village upon village.

Further exploration confirmed Stanley's first impressions of vast natural riches just waiting to be exploited. 'We are banqueting on such sights and odours that few would believe could exist,' he wrote after another trip up river. 'We are like children ignorantly playing with diamonds.'

Leopold had found his colony. Privately he raved about the potential of 'this magnificent African cake'. But he was careful to present the situation in less enthusiastic terms to other European powers, wary of signs of expansionism by the Belgian newcomer. The flag flown at the newly established Congo stations ostensibly belonged to the International African Association, a philanthropic organisation Leopold had set up with the stated aim of wiping out the slave trade and spreading civilisation. Leopold encouraged missionaries to set out for the Congo and at the Berlin conference of 1884–5, at which the world powers carved up Africa, he triggered unanimous applause by proposing the Congo as a free trade zone, open to all merchants. His ambitions for the nation, he said, were purely philanthropic. In return, the Congo Free State was recognised as coming under his personal—as opposed to Belgium's—control.

But, as Marchal's work makes clear, the situation on the ground was to prove rather less high-minded. Clearing the jungle to build roads, stations and—eventually—a railway linking the hinterland with the sea, Stanley's ruthless treatment of his native labourers won him the sobriquet 'Bula Matari' (Breaker of Rocks).

Unable to read the treaties they had signed, local chiefs discovered they had handed over both their land and a monopoly on trade. King Leopold, noted Stanley, in words that could have been used of Mobutu a century later, had the 'enormous voracity to swallow a million of square miles with a gullet that will not take in a herring'.

If the signatures were given 'freely', Stanley left the clan leaders in no doubt that he had the force with which to pursue his interests. He took great delight in demonstrating the wonders of the Krupp canon, the latest in modern weaponry. 'Notwithstanding their professions of incredulity as to its power,' he recounted with satisfaction, 'it was observed that the chiefs took great care to keep at a respectful distance from the Krupp, and when finally the artillerist, after sighting the piece to 2,000 yards, fired it, and the cannon spasmodically recoiled, their bodies also instantaneously developed a convulsive moment, after which they sat stupidly gazing at one another.'

Later on, the Force Publique, a 15,000–19,000-strong army of West African and Congolese mercenaries, was established to ensure Leopold's word became law. Weapons and ammunition poured into the region. Just as Mobutu was later to give the nod to a system of organised looting by instructing his soldiers to 'live off the land', Leopold expected the Force Publique to provide for itself, pillaging surrounding villages in search of food.

Far from being a free trade zone, the colony's very *raison d'être* was to make money for the King. Anxious to attract the foreign capital needed to build railways and bridges, Leopold divided part of the country into concessions held by companies in which he held a 50 per cent stake, with exclusive rights over tracts of forest, ivory, palm oil and mineral wealth. The rest of the country was defined as Crown property, where state agents enjoyed a business monopoly. Independent merchants who ventured into the area in search of ivory found their way physically blocked by Leopold's officials. When the Arab traders operating in the north and eastern reaches of Congo were eventually driven out after a vicious war against the Force Publique, it was not—whatever the Tervuren museum may claim—because of any outrage over their slaving activities, it was because they threatened Leopold's commercial interests.

By then, as the boom in the motor industry escalated Western demand for rubber, Leopold's agents were knowingly mimicking the techniques of the Arab traders that Stanley had decried. Villagers, who had to tap the wild vines growing in the forest for gum, were set cripplingly high production quotas. If they failed to meet the targets, the Force Publique would descend on a village, burn its huts, kill at random and take womenfolk, children or chiefs prisoner until the villagers came to heel. Hostages were used as porters or sold as slaves to rival tribes in exchange for rubber or ivory, and thousands of orphaned children were marched off to Catholic missions to be trained as soldiers for the Force Publique.

Driving the state agents on was a cynical commission system that could double their miserly salaries depending on output and a sliding scale of payment which ensured that those who paid the villagers least for their deliveries of ivory or rubber were rewarded most highly. The lack of compassion seems a little more understandable when one considers the risks inherent in working in the Congo Free State. A staggering one in three state officials desperate enough to try their luck in Africa did not survive their postings, felled by malaria, typhoid or sleeping sickness. With the likelihood of dying in service so high, these young men were none too fastidious about the methods used to ensure output targets were met.

Looking at the mournful black and white photographs taken by appalled missionaries, it is sobering to register that around a century before the amputations carried out by Sierra Leone's rebel forces sent shudders through the West—reinforcing stereotypes of African barbarism—a white-led, European-commanded force had already perfected the art of human mutilation. Soldiers in the Congo were told to account for every cartridge fired, so they hacked off and smoked the hands, feet and private parts of their victims. Body parts were presented to commanders in baskets as proof the soldiers had done their work well. Hence the photographs that, disseminated by the pioneering British journalist Edmund Morel, a precursor of campaigning human rights organisations such as Amnesty International, eventually shocked the outside world into action.

The chicotte, the gallows, mass executions were all liberally applied in a campaign that often seemed to have extermination of races deemed inferior as an incidental aim. The brutality inevitably triggered uprisings. The ferocity of those revolts was glossed over by colonial officers and subsequently downplayed by academics. But Congolese historian Isidore Ndaywel e Nziem records the words of a Captain Vangele, who was attacked four times by canoes manned by tribesmen from Mobutu's own equatorial region, as proof the Congolese were no walkover: 'It was the fiercest battle I have ever experienced in Africa . . . During that fight that lasted nearly three hours, the Yakoma did not cry out once, there was something terrifying about their silence, their cold determination.'

The Force Publique put down the resistance with ruthless effectiveness. Then, as today, no reliable census data existed in the Congo. But as the Force Publique stole children, destroyed families and spread hitherto unfamiliar diseases in its wake, missionaries began to notice an alarming incidence of depopulation taking place. Marchal hesitates to quantify the phenomenon, but Belgian officials were eventually to estimate that the country's population had been halved since the founding of the Congo Free State, implying that 10 million people either died or fled the region. Professor Ndaywel puts the figure even higher, at 13 million.

Leopold had done his best to keep Congo's contacts with the outside world to a minimum, trying to ensure a good press by discouraging visitors and systematically bribing politicians and journalists in Europe. But by the first years of the twentieth century, works such as *Heart of Darkness* were echoing what Roger Casement, a British diplomat, was to officially establish in a 1903 report commissioned by the European powers. Detailing cases of natives being forced to drink white men's urine, having their bound hands beaten till they dropped off, being eaten by maggots while still alive and fed to cannibal tribes on death, Casement destroyed any remaining illusions. What had been laughably dubbed the Congo Free State was an exploitative system premised on forced labour, terror and repression.

Under pressure from foreign allies and his own parliament, the ailing Leopold agreed in 1908, after long negotiations, to hand over Congo to the Belgian state, instead of bequeathing it to his country on his death as he had originally planned. He died a little more than a year later, having never once set foot in the colony his policies had so devastated.

But he had achieved his aim. Congo's massive contribution to Belgium's development is still on show in the capital, if only you know where to direct your gaze. Leopold was a king who wanted to leave his mark on the city of Brussels, and brand it he did, thanks to this independent monetary source he could tap at will.

For visitors interested in the history of Brussels, several companies today offer themed coach trips around the city. A favourite is the Art Nouveau tour, which traces the rise and fall of the design movement that blossomed on the cobbled streets of the hilly city as nowhere else, and the high moment of the tour is undoubtedly the apricot-coloured Hotel Van Eetvelde on Avenue Palmerston, around the corner from the Jamaican embassy and a stone's throw from the plate-glass horrors of Euroland.

Here architect Victor Horta, guiding light of the Art Nouveau movement, was given free reign by Edmond Van Eetvelde, a wealthy diplomat who wanted a fitting venue in which he and his wife could receive business guests. 'I presented him with the most daring plan I had ever, until that point, drawn up,' recalled Horta. Taking advantage of the blank cheque issued him, he produced a building so lavishly decorated, so consistent in its artistic vision, the overall effect is almost nauseating.

From the octagonal drawing hall to the mosaic floors, from the delicate tendrils of the wrought-iron banisters to the motif on the coloured glass roof, the Hotel Van Eetvelde is pure Horta. It is also pure Congo. The hardwoods that lined the ceilings, the marble on the floors, the onyx for the walls and the copper edging each step of the curving staircase all came from the colony. What did not come directly from the colony was paid for with its proceeds, for Van Eetvelde was more than just a well-connected diplomat—he was secretary-general to the Congo. One of Leopold's most trusted collaborators, he was rewarded in 1897 for his loyal services with a baronetcy, before eventually being sidelined by a king whose judgement he had dared to question.

The Hotel Van Eetvelde is only one of the many architectural extravagances Congo's exploited labourers made possible. The Cinquantenaire arch, the grandiose baroque gateway to nowhere, built to celebrate Belgium's golden jubilee; the endless improvements to the Royal Palace at Laeken, including the vast royal greenhouses, Chinese pavilion and Japanese tower; the museum at Tervuren; Ostend's golf course and sea-side arcade and a host of other works were all provided by the Congo. But there was more, much more, and not all of it quite so obvious to public eyes: presents for Leopold's demanding young mistress; a special landing stage for the yacht he, like Mobutu later, would use as a place to hide away from an increasingly hostile public, spending sometimes months aboard; Parisian châteaux; estates in the south of France and a fabulous villa in Cap Ferrat, not far from where Mobutu would buy a mansion.

The two men shared more than just a knack for large-scale extortion and lavish spending tastes. Indeed, in money matters, the present echoes the past to an almost uncanny extent. Both leaders were to prove remarkably adept at squeezing loans out of gullible creditors and luring private investors with a taste for adventure to Africa. Both covered their tracks with a system of fraudulent book-keeping. Both indulged in similar stratagems in an attempt to cheat the taxman after their deaths and both, having feathered their own nests, left Congo with a heavy burden of debts to be settled after they quit the scene.

In contrast to most African colonies, the Congo Free State was a money-maker almost from birth, thanks to Leopold's eye on the bottom line. But the king did his best to conceal that fact, succeeding so well in obscuring the true situation that a British journal of the day erroneously reported: 'It is by no means certain that Belgium will not tire of the Congo. Already this vast area has been a huge disappointment to the mother country. Its resources and population have not proved in any way equal to Mr Stanley's florid accounts.'

Pleading near bankruptcy, Leopold managed to win two major loans worth a total of 32 million francs from the Belgian state in 1890 and 1895, paid out in yearly instalments. But while the faithful Van Eetvelde was drawing up fictitious budgets underestimating revenues, thereby ensuring the government maintained subsidies for a colony the public had never wanted in the first place, profitability was sharply on the rise. By 1901 ivory exports stood at 289,900 kilograms and rubber production had gone from 350 to 6,000 tonnes a year. Congo was providing more than a tenth of world production of this key raw material, bringing in somewhere between 40 and 50 million francs a year. The king also made money by issuing more than 100 million francs worth of Congo bonds, effectively printing money with the same liberality as Kinshasa's central bank was later to show when it came to issuing notes.

When Leopold was finally forced to hand the colony over to Belgium, he did so at a high price, wheedling 50 million francs from the government in recognition of his endeavours. The Belgian government, which had always been assured it would never be sucked into the king's African adventures, found itself agreeing to assume Congo's 110 million francs in debts—much of that sum comprising the bonds Leopold had issued—and contribute nearly half as much again to completing the building projects the king had drawn up in Belgium.

No one will ever know for certain how much profit Leopold himself drew from the Congo Free State. He adopted the methods beloved of many a modern-day African strongman when it came to trying to hide the extent of the wealth he had accumulated. Real estate was bought through aides, money secretly funnelled into a foundation dedicated to building projects, and shadowy holding companies set up in Belgium, France and Germany. Before handing over responsibility for his African colony, Leopold was careful to burn much of the Congo documentation, protecting himself as far as he could from the scrutiny of future scholars. Belgian investigators only succeeded in unravelling the complex network of his investments in 1923.

By then the world's attention had moved elsewhere, satisfied that the human rights abuses in Congo had halted with the Belgian's government takeover. Not so, insisted Marchal, who aimed to challenge this comfortable myth in the book he was currently writing about the system of forced labour imposed by Belgium's Union Minière, the company that continued running the mines in Congo's southern Katanga region well after independence. 'When I finished writing about Leopold, I thought it would be over for me, because I believed all those professors who said when Belgium took over everything was wonderful. But I've seen that things remained the same, the system was nearly as brutal, it just became more hypocritical. I now have material for another three or four books.'

Marchal's own memories might have suggested as much. The system of forced cultivation in the cotton industry he enforced as a young man lasted until independence in 1960; use of the chicotte, that mainstay of colonial rule, was outlawed only ten months before Belgium pulled out. The officials who had worked under Leopold had a new master but largely remained in situ. Reforms were applied only slowly. It was only after the Second World War, Marchal now believed, that the Belgian Congo became 'a colony like the others'.

Even then, Belgium hardly distinguished itself. True, it had established an infrastructure whose modernity was marvelled at by European visitors. To take just one example, Congo at independence had more hospital beds than all other black African countries combined. But daily life resembled that adopted in South Africa under apartheid rule.

The capital was divided into the indigenous quarters and the Western zone, where blacks were not allowed after a certain time and would be refused drinks in hotels and restaurants which were reserved for whites only. Referred to as 'macaques' (monkeys)—a term still contemptuously spat out by heavy-drinking expatriates in Kinshasa—Congolese were set the qualification of 'évolué' as a target. This was a certificate indicating they were Africans who had 'evolved' far enough to adopt European attitudes and behaviour. But it was not enough to allow them to accede to positions of responsibility and power.

Certain experiences are calculated to stick in the gullet. Long, long after independence, one of the MPR's leading lights would sometimes recall the time when a Belgian colonial official came round to verify the cleanliness of his parents' toilet before issuing the permit that allowed them to buy wine. In schools, children from such 'evolved' Congolese families would be taken aside each week to be checked for fleas, an indignity spared their white classmates.

Acting on the principle of 'pas d'élites, pas d'ennemis',—the theory that an educated African middle class would prove dangerously subversive—the Belgians did virtually nothing to pave the way for independence, expected in 1955 to be decades off. When the government was forced to hand over in the face of growing protests in 1960, only seventeen Congolese youths had received a university education. The withdrawal was one of the most abrupt in African history.

Why did this small European nation prove such an appalling colonial power? One gets the impression that Leopold was rushing so desperately to catch up with his foreign allies, self-restraint and principles were simply jettisoned along the way. Maybe a country in its infancy did not possess the self-confidence necessary to show magnanimity when imposing nationhood on others. As tribally divided as the nations hacked arbitrarily from Africa's land mass by the colonisers, Belgium barely had a sense of itself, let alone itself in the novel role of master.

Marchal, convinced modern Belgium owed the Congolese some kind of reparation in recognition of its errors, even if it only took the form of a more relaxed visa system, seemed to lay the blame on a failure of imagination. A 'small country with small horizons', as Leopold himself contemptuously described it, Belgium regarded the Congo as a money-making opportunity, and little else, unlike colonial nations with longer imperial traditions behind them and loftier ideals.

One former ambassador—not a Belgian—put it rather more bluntly: 'The Belgians were awful in Congo because they had no grandeur themselves. This was the Zaire of Europe, a ratty little country divided amongst itself, and it proved incapable of aspiring to the heights.'

Not long ago, strange notices began appearing over the clothes racks in the slick designer shops and perfumeries lining Boulevard de Waterloo, the broad thoroughfare that carves an ugly swathe through the heart of Brussels.

They were written in Lingala, a language incomprehensible to most Belgians. They warned their readers anyone caught stealing would not only be arrested and charged, but expelled from Belgium and sent back to their country of origin. Their appearance, somewhat at odds with the fur-coated, poodle-carrying sophistication of this most European of cities, was a tribute to the effectiveness of the Congolese women hit-squads who had taken to systematically shoplifting designer labels in the area.

'It's time to repay the colonial debt. On va kobeta' ('We're going on a raid'), the women would say, as, with the rumbustious energy only an African market trader can bring to her task, they set off in search of Versace and Yamamoto jackets, Gianfranco Ferre and Jean-Paul Gaultier slacks, Kenzo accessories and Church shoes—anything decreed cool by the trendsetters of the day.

The designer shops had only themselves to blame. They were, after all, displaying their goods within temptingly easy striking distance of the poor Congolese ghetto that nestles compactly in the covered galleries and cobbled streets of Ixelles, just off the Porte de Namur. Few districts in the Belgian capital can rival 'Matonge', focal point for the Congolese community, when it comes to juxtaposing inordinate personal vanity with the chronic inability to meet the cost of a heightened sense of style.

Nicknamed after Kinshasa's heaving popular quarters, because, like its namesake back home, this is a district where 'ça bouge' (things move), Matonge is like a long draught of Congolese essence that has been decanted and boiled down to its purest concentrate. There is something brave, almost foolhardy, about the way this tiny ghetto turns its back on the Belgian present of tramlines, dark streets and narrow houses to recreate a more familiar reality.

In the hairdressers—and every second shop seems to be a hairdresser, its window crammed with wigs and hair extensions—Congolese women have their hair straightened or young blades chat. The greengrocers here sell fat stalks of sugar cane, nobbly sweet potatoes, heaps of the greens used to make pondu, the Congolese alternative to spinach, deadly red chillies and small, pale green aubergines. The front pages of Congolese newspapers, *Le Soft, Le Palmarès, Le Phare*—with all their tunnel-vision, their obsession with the domestic political scene—are stuck against café windows; 'waxes', the bright Dutch prints used to make women's wraps, lie folded on display in neat rows and even the gold on sale in the jewellers has that pinkish tinge associated with Africa.

Restaurants serve chicken in peanut sauce, fish wrapped in palm leaves and it is even possible to find such delicacies as caterpillar, crocodile—the oysters and caviar of Kinshasa's culinary scene—or chikwange, the leaf-wrapped blocs of fermenting cassava paste that, to the uninitiated, resemble nothing quite so much as warm carpet glue.

In the old days, a tailor here turned out the awkward abacost jackets made obligatory by Mobutu. The ghetto even has its own radio station. Broadcasting from an abandoned military barracks, Radio Panik feeds its listeners a diet of Koffi Olomide, Zaiko Langa Langa, Papa Wemba, or whoever dominates the Congolese music scene of the day, plus, most crucially for a public hungry for information from home, a weekly résumé of Congolese news.

Sitting squat in the city centre, Matonge is a psychological world away from the leafy suburbs of Rhode St Genesè, Uccle and Waterloo to the south of Brussels, where Mobutu's former aides live in marble-floored mansions, over garages where the Mercedes is parked alongside the BMW. Just as the presence of Mobutu's château in Brussels's chic suburbs acted as a magnet for the Congolese elite, who set up their court around the big man, Matonge, at the other end of the social scale, owes its existence to the Maison Africaine, a hostel where those shaking the red dust of the continent from their feet could stay for next to nothing, often lingering for years on end.

Cafés sprang up serving the food homesick new arrivals missed, as did music shops and the nightclubs, Le Mambo, La Référence, Hollywood City, which only come alive in the early hours. Matonge became an area the 15,000 Congolese living, studying and working in Belgium recognised as a second home, a place where the Congolese genius for finding creative solutions to the problems of existence surfaced.

Family in dire straits at home? There are agencies here where you can go, deposit 100 dollars, sure in the knowledge that a dependant at the other end in Kinshasa will receive another 100-dollar bill, all without going through a bank. Relatives going hungry or can't afford the price of an electrical appliance? The same procedure is available for a sack of rice or a fridge. And when disaster really strikes you can even, through these tiny offices, arrange a funeral back in Congo.

The entrepreneurship extends well beyond the law's reach. A vibrant trade in second-hand cars, drugs and forged cheques, prostitution and fake visas, plus the designer brand shoplifting, has prompted Belgium's police to establish a unit specialising solely in crime committed by members of the Congolese community, something of a mark of distinction given the far greater numbers of Moroccans and Turks in Brussels.

Despite all the cheering inventiveness, there's a tragic poignancy about Matonge. The alliterative Lingala slang residents use to refer to life abroad is premised on vaunting ambition, but the aspirations come tinged with a sense of inferiority. For those abandoning Kinshasa, despairingly dubbed 'Kosovo', Belgium is 'lola', or 'paradise'. Paris, another favourite destination, is known as 'Panama'. Europe is 'mikili', 'the promised land', inhabited, appropriately enough, by 'mwana Maria', 'the children of the Virgin Mary'—whites.

This is a community determined to outstay its welcome, made up of forty-year-old students with a smattering of children and fistfuls of degrees; of young men playing up their brushes with the law in Kinshasa in the hope of winning the sobriquet of 'political asylum-seeker'; of youths plotting marriages of convenience with Belgian mates: all and any methods are acceptable in the quest for the ultimate prize—a permit allowing an indefinite stay in Europe.

When it is won, such documentation rarely goes to waste. 'Whites say that all blacks look alike,' explained Leon, a philosophy graduate studying accountancy, 'so someone with papers will lend them to a friend who wants to cross into France or Switzerland, who will then post them back to Brussels.' Without the paperwork, work outside the informal sector is impossible. So Brussels's restaurant kitchens, its building sites, its minicab firms, are staffed by Africa's most well-qualified students.

The sense that only the West offers hope of improvement is enough to make even the uninspiring seem acceptable. 'I have friends who are vegetating here. They do nothing, they stagnate, but they don't dare go back,' said Leon. 'In the eyes of their families, returning from Europe means they have failed. And the worst thing you can have happen to you, the most humiliating, is to be expelled.'

Other African communities forced into exile organise guerrilla campaigns from abroad, hatch plots, or draw up political programmes for the distant day when they hope to take power. For decades, Eritrean émigrés ran an efficient informal tithing system which funded the rebel movement that eventually pushed Ethiopian occupiers out of their territory. Despite boasting one of the continent's most formidable dictators as an antagonist to rally against, the Congolese have nothing to match this. If a rebel campaign is being fought in the east of their country, amongst the young men of Matonge there is no talk of donning camouflage and signing up. The biggest opposition party had closed its offices 'for security reasons', I was told, but administrative incompetence was more likely to be the cause. The collective sense is missing.

Congolese themselves acknowledge the lack, with a shrug of the shoulders and the rueful honesty that is in itself part of the problem of proscribed ambitions and low expectations. Each man's aim is to leave Congo, acquire qualifications, and build a life somewhere else. Let someone else draw up a constitution. Let someone else rebuild the country. Experience has taught that politics is a game played by conmen and hypocrites.

What adds a bitter edge to this undignified scramble for the exit is the realisation that while thousands of Congolese immigrants would not be living in Brussels, Antwerp, Ghent and Liège were it not for their country's historical ties with Belgium, a younger generation of Belgians is virtually unaware of that painful colonial past.

'There is no African memory left,' acknowledges Marcelin, who works for a struggling Congolese state company with offices in Brussels. 'There are very few Belgians left in parliament or the ministries who worked in the colony, so the sentimental attitude of the past has gone. All that is left is a sense of disappointment with our leaders and negative associations of disaster, death and dictatorship. Young Belgians assume Congolese either make music all the time or are petty crooks. There is no sense of responsibility for what their country did in the Congo, let alone guilt.'

Despite the intimate historical relationship, no Belgian newspaper or radio station has a foreign correspondent permanently based in Kinshasa. In a country struggling with its own contradictions, preoccupied with prickly Francophone-Flemish relations, Belgian colonial history is not taught at school. The distorted vision of history the Royal Museum at Tervuren set out to sanctify has been incidentally fostered by the political sensitivities of modern Belgium.

Young Bruxellois live in a city dotted with baroque monuments funded with the proceeds of the Congolese state, scattered with antique shops selling Congolese masks and home to the biggest community of Congolese living abroad. Yet *King Leopold's Ghost*, the first book in years to stir a general debate on the topic, was written by an American, not a Belgian.

As Jean Stengers, a retired professor who has written copiously on the Congo Free State, freely admitted, his pet subject remains almost exclusively in the narrow intellectual domain, a closed book to most fellow nationals. Working from a study crammed with leather-bound volumes and papers looking out on the bleak Rue de Couronne, the white-haired academic had criticised Marchal for his interpretation of history, arguing that the former diplomat ignored the fact that national glorification, rather than personal enrichment, was Leopold's prime motivating factor. But if they differed in their views of the king, the two men shared a rueful awareness the topic they both regarded as of such importance was a matter of general indifference.

What feelings existed, Stengers said, were amongst a disappearing generation and—astonishingly—they were scarcely feelings of shame. 'In the older generation, many of whom served in the

Congo, the strongest feeling is one of injustice done. There's a deep sense that magnificent things were given to the Congolese and we were rewarded with huge ingratitude. But the public at large has lost interest in the Congo. For the new generation, ignorance of Belgian history is nearly as great as ignorance of Congo's history.'

Knowing nothing about the past, of course, frees a population from any sense of blame for the present. How convenient was all this forgetting, I wondered as I walked down the steps of Stengers' house, given the débâcle of modern-day Congo?

The question Belgian researchers into the Congo Free State hate to be asked is whether there is any causal link between Belgium's exploitative regime and the excesses of Mobutu's rule, whether a frighteningly efficient kleptocratic system effectively softened up a community for a repeat performance.

Marchal had brushed it anxiously away, pleading that he was a historian rather than an intellectual, and it was not for him to make such judgements. When put to Professor Stengers, the question had been rejected with a categorical shake of the head. Citing sociological studies conducted in the Great Lakes region, he said what was striking was the lack of memories of the Leopold era amongst the local population. So how could there be any causal link?

But that, I thought, seemed to be missing the point. Plunging into the dreadful detail of Leopold's reign, I, too, had been surprised by how few of these horrors—surely the stuff of family legends passed down from patriarch to grandson—had ever been mentioned to me by Zairean friends. But it wasn't necessary to be an expert on sexual abuse to know it was possible to be traumatised without knowing why; that, indeed, amnesia—whether individual or collective—could sometimes be the only way of dealing with horror, that human behaviour could be altered forever without the cause being openly acknowledged.

In Belgium I began to sense the logic behind many of the peculiarities that had puzzled me living in Kinshasa, a city where everyone seemed to complain about how awful things were but no one seemed ready to try changing the status quo; where grab-it-and-run was the principle of the day and long-term planning alien. Page after page, the picture painted by Marchal had struck a chord.

Coming after the raids of the hated Force Publique and the slave traders, Mobutu's looting soldiers were just more of the same. After the crippling production targets set by Leopold's agents, the informal 'taxes' levied by corrupt officials must have seemed benevolent in comparison. Having seen their revolts against the Belgian system crushed by troops wielding such horrors as the Krupp cannon, who still had the courage to rise up against Mobutu's army, however shambolic it came to seem to Western eyes? And how could the Congolese ever value or build on an infrastructure and administration imposed from above, using their sweat and blood as its raw materials?

Keep your head down, think small, look after yourself: these constituted the lessons of Leopold. The spirit, once comprehensively crushed, does not recover easily. For seventy-five years, from 1885 to 1960, Congo's population had marinated in humiliation. No malevolent witch-doctor could have devised a better preparation for the coming of a second Great Dictator.

CHAPTER THREE

Birth of the Leopard

'Politics are too serious a matter to be left to the politicians.'

—Charles de Gaulle

There was a moment in 1960, when, if a white man had stayed his hand and decided not to get involved, the newly independent Congo's history would have taken a very different course. It was the split second when a young CIA station chief who had crossed a tense capital walked around a corner at one of Leopoldville's military camps and surprised a man in civilian clothing taking aim at a figure walking away.

'I guess I was a Boy Scout too long, because without thinking I jumped at the man with the pistol. Then I was sorry, because it turned out he was very strong,' he recalled. 'We rolled around in the dirt and I finally remembered something I'd learnt in army training. He had his hand in the trigger guard and I pulled it back until the bone snapped.' The scuffle attracted the attention of the intended victim's bodyguards who, misunderstanding the situation, promptly started beating up the Good Samaritan. 'All I could think about,' he chuckled, 'was why the hell did I get involved?'

A generation of Zaireans might today ask themselves the very same question, but with a greater degree of asperity and rather less humour. For the target of the botched assassination attempt, staged at the orders of an aspiring Congolese politician with Soviet contacts, was Colonel Joseph Désiré Mobutu, who had just taken over the running of the country. If the white man in question—Larry Devlin—had not intervened, who knows what route the country would have followed?

But then, interference, whether muscular or subtle, was always something of a forte of Mr Devlin's. His role in the traumatic events of Congo's post-independence period was to leave him one of the most notorious CIA men in history, an example of just how far the United States was willing to go in that epoch to sabotage the Soviet Union's plans for global communist expansion.

Mr Devlin's life had been one of commotion: a *bête noire* for a generation of Africans still fuming over the way superpower intervention dictated events on the continent during the Cold War, he had been accused by conspiracy theorists of engineering the murder of Patrice Lumumba—Congo's first, inspirational prime minister. Grown fragile and snowy-haired in his seventies, he had survived wars (two), uprisings (two), crash landings (four), heart attacks (several), beatings and assassination attempts (many) and a medical death sentence (two months to live, delivered, mistakenly, in 1984 when doctors spotted what they thought was an inoperable brain tumour).

It had not all been pain and suffering. He learned to dance in Leopoldville's sweaty nightclubs, argued politics into the small hours with the young men who were to become Congo's movers and shakers and got tipsy on the sun-baked sandbanks of the Congo river.

But it had all taken its toll, leaving him unsteady on his feet, floating above the pavement with the uncertain grace of a fifteenth-century schooner setting out on its first journey to the New World, an old-fashioned gentleman who opened car doors for a lady, gently insisted on paying and who dressed with a studied elegance wholly appropriate for a man who once, during some bizarre career interlude, ghosted articles for French fashion designer Jacques Fath.

The consultancy work Devlin continued doing on Africa from his home in Virginia did not take up all his time and in retirement he had grown chatty. Two instincts were warring within him. On the one hand, he had been attacked too many times by the press as the kingmaker who put Mobutu in power, starred as the ruthless secret agent in too many thinly fictionalised accounts of the Congo crisis, not to be wary. On the other hand, with time on his hands and as the kind of man who clearly enjoyed female company, this was a not entirely unpleasant opportunity to set the record straight.

His voice had the gravelly timbre of a man who smoked three packets of cigarettes a day until a brush with open-heart surgery. His hands—creased by a million experiences, the wedding ring so deep-set in the flesh it seemed welded to the bone—would give a palm-reader pause for thought. But the brain was as keen and irreverent as ever. And with his defiant insistence that he regretted nothing about the CIA's support for Mobutu, Larry Devlin was a reminder that whatever happened in the end, there was a time when Mobutu was not just the hope of interfering Americans obsessed with domino metaphors, but of a population exasperated by the dithering, squabbling and tribalism of its civilian leaders.

'What you must never forget is that there were many periods to Mobutu. You saw the pitiful end. But he was so different at the start. I can remember him as a dynamic, idealistic young man who was determined to have an independent state in the Congo and really seemed to believe in all the things Africa's leaders then stood for.'

They first met in Brussels in early 1960, when members of Congo's embryonic political establishment found themselves negotiating independence terms with their colonial master. Five years earlier, a Belgian expert had triggered an uproar at home by putting forward a thirty-year programme for a pull-out. Most Belgians believed they had another 100 years to go, plenty of time to train up and educate their eventual replacements. Subsequent events had exposed how out of touch even that supposedly accelerated schedule really was: riots in Congo's major cities, increasingly vocal demands by the country's 'evolués' and France's and Britain's disengagement from their own African possessions had forced Belgium to realise decolonisation was due.

Having accepted the principle, Brussels set about formalising its withdrawal with indecent haste. But while Belgium was pulling up the colonial drawbridge, other powers were becoming interested in the new opportunities the postwar configuration was throwing up. The two sessions of round-table talks in Brussels provided a rare chance for their representatives to size up the future leadership of the Congo, whose size, geographical position and huge resource base made it the natural linchpin of central Africa.

Devlin was working in Brussels at the time. He was a young man who already had a lifetime's experience behind him. A committed anti-Nazi, he had interrupted his college studies to sign up as a private in the US Army, had served in Italy and been injured. Returning to college, he had been recruited by a Central Intelligence Agency no doubt impressed by his war record, his sharp mind and his mastery of several languages. His speciality was Soviet operations and he had become skilled at 'turning' Soviet bloc officials, a process he remembered now as being 'better than an orgasm' when successfully pulled off.

But he had angered a superior in the process and his career had fallen into something of a slump when the Congolese negotiations opened and he began picking up alarming signs of Soviet activity in Brussels: 'I noticed that Soviets were contacting one by one every member of the various delegations at the round table conference. I got curious as to what they were doing and why. What I found was that they were essentially spotting, assessing and trying to recruit. It was a classic effort on their part. The Russians wanted to use the Congo as their stepping stone into Africa.'

The Soviets knew they had a potential ally in Patrice Lumumba. A public speaker with a near-miraculous ability to win round his audience, this former post office employee had become the spearhead of Congo's independence campaign. Inspired by the pan-Africanism of Ghana's Kwame Nkrumah and Guinea's Sekou Touré, he was a flamboyant, erratic figure, bubbling with ideas. Released from jail to attend the Brussels meeting, he was brimming with resentment over Western imperialism in Africa.

The Soviet contacts with the delegations from Leopoldville were enough to ensure the US embassy in Brussels got involved. The American ambassador threw a reception for the Congolese and Devlin and his embassy colleagues launched themselves in a very deliberate bout of networking. 'Each of us drew up a list of 10 or 12 people we had to meet and afterwards we all got together to discuss our impressions. One name kept coming up. But it wasn't on anyone's list because he wasn't an official delegation member, he was Lumumba's secretary. But everyone agreed that this was an extremely intelligent man, very young, perhaps immature, but a man with great potential. They were right, because that was Mobutu.'

The next time Devlin met Mobutu was in the Congo Republic—his new posting—as all hell broke loose. Less than a week after independence on 30 June 1960, Belgium's haste was having inevitable consequences. Told there were to be no immediate moves to 'Africanise' an army exclusively commanded by Belgian officers, Congo's troops mutinied, whites were beaten and raped and the Belgian technicians who ran the country's administration headed *en masse* for the airport.

Prime Minister Lumumba appointed Mobutu army chief of staff. Touring the country's military bases, playing up his own army experience, Mobutu persuaded the soldiers to return to barracks. But the mutiny was not Lumumba's only problem. Belgian paratroopers had landed in what the Congolese assumed to be a second colonial takeover. The new state seemed doomed to break up as, encouraged by a former colonial master bent on ensuring continued access to Congo's mineral wealth, first copper-producing Katanga and then diamond-rich Kasai seceded.

The UN responded to the crisis with extraordinary speed. Its reaction time, like the hordes of journalists who flooded into Congo to cover those years, was a measure of the enormous hopes the West was pinning on Africa during those years. Impossible as it is to imagine in the year 2000, when the renewed threat of national fragmentation raises barely a flicker of international interest, the Congo of the 1960s was one of the world's biggest news stories.

The first UN troops landed in Leopoldville the day after Lumumba and President Joseph Kasavubu called on the UN Security Council for protection from foreign aggression. But Lumumba, who had hoped they would help snuff out the secession movements in the south, was bitterly disappointed by their limited mandate, which barred them from interfering in Congo's internal conflicts.

Feeling betrayed by the West, Lumumba turned to the Soviet Union for help, requesting transport planes, trucks and weapons to wipe out the breakaway movements in Kasai and Katanga. Nikita Khrushchev obliged. The military aid arrived too late to prevent a bloody débâcle in Kasai, where the Congolese army lost control, slaughtering hundreds of Luba tribespeople. But for Washington what mattered was that this was the first time Moscow had intervened militarily in a conflict so far from its own borders. It represented a dangerous ratcheting up of the Cold War game.

'I had a little Congolese sitting at the airport counting any white man who came off a Soviet aircraft in batches of five. Roughly 1,000 came in during a period of six weeks. They were there as "conseillers techniques" and they were posted to all the ministries,' recalled Devlin. 'To my mind it was clearly an effort to take over. It made good sense when you stopped to think about it. All nine countries surrounding the Congo had their problems. If the Soviets could have gotten control of the Congo they could have used it as a base, bringing in Africans, training them in sabotage and military skills and sending them home to do their duty. I determined to try and block that.'

It was a line of argument that was to justify more than three decades of American support. But if for Washington Lumumba was showing a worrying resemblance to Fidel Castro, Devlin himself, ironically enough, never believed in the sincerity of Lumumba's conversion to the Soviet cause. 'Poor Lumumba. He was no communist. He was just a poor jerk who thought "I can use these people". I'd seen that happen in Eastern Europe. It didn't work very well for them, and it didn't work for him.'

The wave of Soviet arrivals triggered the collapse of Lumumba's strained relations with Kasavubu, Congo's lethargic president. At times, too many times, politics in Congo resembled one of those hysterical farces in which policemen with floppy truncheons and red noses bounce from one outraged prima donna to another. 'I'm the head of state. Arrest that man!' 'No, I'M the head of state. That man is an impostor. Arrest him!' Only the reality was more dangerous than amusing. In a surreal sequence the prime minister and president announced over the radio that they had sacked each other. Mobutu was put in an impossible position, with both men ordering him to take their rival into custody.

The army chief of staff was already unhappy with the turn events were taking. 'The Russians were brutally stupid. It was so obvious what they were doing,' marvelled Devlin. 'They sent these people to lecture the army. It was the crudest of propaganda, 1920s Marxism, printed in Ghana in English, which the Congolese didn't understand. Mobutu went to Lumumba and said "let's keep these people out of the army". Lumumba said "sure, sure I'll take care of that", but he didn't. It kept happening and finally Mobutu said: "I didn't fight the Belgians to then have my country colonised a second time." '

Exactly what role Devlin played in determining subsequent events was not clear. Cable traffic between Leopoldville and Washington shows he received authorisation for an operation aimed at 'replacing Lumumba with a pro-Western group' in mid-August 1960. Despite his friendliness, Devlin remained bound by the promises of confidentiality made to the CIA, contemptuous of those in the intelligence services who leaked government secrets. All he would say was that it was during those dramatic days that he really got to know Mobutu. The army chief was already being leaned on by the Western embassies—whose advice was given added weight by the fact that they were helping him pay his fractious troops—President Kasavubu, the student body and his own men. No doubt the CIA station chief brought his own persuasive skills, that talent acquired during years of 'turning' Soviet personnel, into play as Mobutu edged towards one of the hardest decisions of his life.

The eventual outcome, Devlin acknowledged, came as no surprise. On 14 September 1960, Mobutu neutralised both Kasavubu and Lumumba in what he described as a 'peaceful revolution' aimed at giving the civilian politicians a chance to calm down and settle their differences. Soviet bloc diplomatic personnel were given forty-eight hours to leave. The huge African domino had not fallen: Congo had been kept safely out of Soviet hands.

It was exactly what Washington wanted. But Devlin nonetheless rejected any notion of Mobutu being an American tool. 'He was never a puppet. When he felt it was against the interests of the Congo, he wouldn't do it, when it didn't go against his country's interests, he would go along with our views. He was always independent, it just happened that at a certain point we were going in the same direction.' And like many commentators of the day, he still believed that Mobutu, an earnest twenty-nine-year-old pushed to prominence by a failure of leadership and a jumble of cascading events rather than personal ambition, was genuinely reluctant to take over in 1960. Such modesty would not last very long.

Who was the man who so impressed Devlin and the diplomats as they circulated, glasses in hand and mental notebooks at the ready, at the reception in Brussels?

Joseph Désiré Mobutu was born on 14 October 1930 in the central town of Lisala, where the Congo river runs deep and wide after its grandiose circular sweep across half a continent. That early proximity to the river, he always claimed, left him with a visceral love of the water. 'I can say that I was born on the river . . . Whenever I can, I live on the river, which for me represents the majesty of my country.'

He was a member of the Ngbandi tribe, one of the smaller of the country's 200-plus ethnic groupings. Anthropologists believe the Ngbandi trace their lineage back to the central Sudanese regions of Darfur and Kordofan, an area that was repeatedly targeted by Moslem Arab conquerors from the sixteenth century onwards.

Fleeing the slave raids and Islamicisation, his animist ancestors fled south, heading for the very equatorial heart of the continent, where they in turn subjugated the local Bantus. Safe in the glowering forests that later so terrified Western explorers, they intermarried and the Ngbandi—who took their name from a legendary fighter—gradually acquired an identity. They emerged as a loose affiliation of war-like tribes speaking the same language and straddling the Ubangi, a subsidiary of the great Congo river, with one foot in what is today Central African Republic and another in Congo.

Like all autocrats, Mobutu was later to mythologise his own upbringing. In one story, almost certainly apocryphal, he described walking in the woods with his grandfather. When a leopard leaped from the undergrowth, the boy shrank away. The grandfather remonstrated with him and, ashamed and piqued, the young Mobutu seized a spear and slew the leopard. 'From that day on,' said Mobutu, 'I am afraid of nothing.' He was to use the animal at the centre of this coming-of-age fable as his personal insignia, a symbol of pride, strength and courage. It was also the origin of his trademark leopard-skin hats which, in a curious juxtaposition of machismo and decadence, he had made by a Paris couturier, keeping a collection of at least seven on hand.

The truth of those early years is somewhat less romantic. Some of Mobutu's contemporaries recall that in the pre-independence era, there was a tendency amongst city dwellers to sneer at the Ngbandi, marooned in one of the least accessible zones of Africa, as coarse rustics who had barely shed their loin-cloths in favour of Western-style clothing; good hunters, yes, but in need of some urban refinement.

Mobutu would later ensure that changed. But when he was growing up, he belonged to a tribe regarded as 'sous-evolué'—under-evolved. He shared with many prominent men a keen awareness of his humble origins, a source of resentment pushing him ceaselessly, fruitlessly, to try and prove his superiority. And if Mobutu's ethnic origins were not enough of a burden, there was another issue calculated to niggle at the confidence of an impressionable youngster—his parentage.

His mother Marie Madeleine Yemo, whom he adored, was a woman who had notched up her fair share of experiences. She had already had two children by one relationship when her aunt, whose marriage to a village chief was childless, arranged for her niece to join her husband's harem. It was a kind of brood-mare, stand-in arrangement that, while strictly in accordance with local custom, must have contained its share of bitterness and humiliation for both of the women concerned.

Mama Yemo, as she was eventually to be known to the nation, bore the chief two children, then twins who died. Suspecting her aunt of witchcraft, she fled on foot to Lisala. It was there that she met Albéric Gbemani, a cook working for a Belgian judge. The two staged a church wedding just in time, two months before Joseph Désiré Mobutu's birth. The boy's name, with its warrior connotations, came from an uncle.

Recalling his youth, Mobutu later had more to say about the kindness shown by the judge's wife, who took a shine to him and taught him to read, write and speak fluent French, than his own father, who barely features. 'She adopted me, in a way. You should see it in its historical context: a white woman, a Belgian woman, holding the hand of a little black boy, the son of her cook, in the road, in the shops, in company. It was exceptional.'

Given that Albéric died when Mobutu was barely eight years old, the dearth of detail about his father is perhaps not surprising. But that lacuna was later seized upon by Mobutu's critics, who would caricature their leader as the bastard offspring of a woman only a few steps up from a professional prostitute.

With his mother relying on the generosity of relatives to support her four children, Mobutu's existence became peripatetic as she moved around the country. Periods in which he ran wild, helping out in the fields, alternated with stints at mission schools. He later claimed that religious exposure left him a devout Catholic, but as with many Congolese, his Christianity never ruled out a belief in the African spirit world which left him profoundly dependent on the advice of marabouts (witch-doctors).

Mobutu finally settled with an uncle in the town of Coquilhatville (modern-day Mbandaka), an expanding colonial administrative centre. The placing by rural families of their excess offspring with urban relatives who are then expected to shoulder their upkeep and education for years, often decades, is extraordinarily prevalent in Africa. Puzzling to Westerners, such generosity is a manifestation of the extended family which ensures that one individual's success is shared as widely as possible. But the burden is often almost too heavy to bear, and such children never have it easy. For Mobutu, life was tough. Perhaps the austerity of those days, when he depended on a relative for food and clothing, explains his love of excess, the unrestrained appetites he showed in later life.

In Coquilhatville he attended a school run by white priests, and the child whose precocity had already been encouraged by a white woman began to acquire a high profile. Physically, he was always big for his age, a natural athlete who excelled at sports. But he wanted to dominate in other ways as well. 'He was very good at school, he was always in the top three,' remembers a fellow pupil who used to play football with Mobutu in the school yard. 'But he was also one of the troublemakers. He was the noisiest of all the pupils. The walls between classrooms were of glass, so we could see what was going on next door. He was always stirring things up. It wasn't done out of malice, it was done to make people laugh.'

One favourite trick was making fun of the clumsy French spoken by the Belgian priests, most of whom were Flemish. 'When they made a mistake he would leap up and point it out and the whole room would explode into uproar,' said a contemporary. Another jape involved flicking ink darts at the priest's back while he worked at the blackboard, a trick calculated to get the class giggling.

In later life, like any anxious middle-class parent, Mobutu would drum into his children the importance of a formal education. One such lecture occurred when the presidential family was aboard the presidential yacht, moored not far from Mbandaka. On a whim, Mobutu sent for the priests from his old school and ordered them to bring his school reports. Miraculously, they still had them and

Nzanga, one of Mobutu's sons, remembered his father proudly showing his sceptical offspring that, academically at least, he had been no slouch.

Given that he did well academically, Mobutu, known as 'Jeff' to his friends, was forgiven a certain amount of unruliness. But the last straw came in 1949 when the school rebel stowed aboard a boat heading for Leopoldville, the capital of music, bars and women regarded by the priests as 'sin city'. Mobutu met a girl and, swept away by his first significant sexual experience, extended his stay. After several weeks had passed, the priests asked a fellow pupil, Eketebi Mondjolomba, where Mobutu had gone.

'Since we lived on the same street, I was supposed to know where he was and I said, in all innocence, he'd gone to Kinshasa,' remembered Eketebi, who was still grateful that Mobutu later laughingly forgave—while definitely not forgetting—this youthful indiscretion. 'At the end of the year, that was one of the reasons why he was sent to the Force Publique. It was the punishment the priests and local chiefs always reserved for the troublesome, stubborn boys.'

The sudden expulsion was a shock. It meant a seven-year obligatory apprenticeship in an armed force still tainted by a reputation for brutality acquired during the worst excesses of the Leopold era. But for Mobutu the Force Publique was to prove a godsend. Here the natural rebel found discipline and a surrogate father figure in the shape of Sergeant Joseph Bobozo, a stern but affectionate mentor. In later life, bloated by good living and corroded by distrust for those around him, he would wax nostalgic about the austere routines of army life and the simple camaraderie of the barracks. Looking back, he recognised this as the happiest period of his life.

In truth, Mobutu was never quite as much of a military man as he liked to make out. Of more importance in furnishing his mental landscape was the fact that he managed to keep his education going in the Force Publique, corresponding regularly with the mission pupils he had left behind, who kept him closely informed of how their studies were progressing. On sentinel duty, carrying out his chores, he read voraciously, working through the European newspapers received by the Belgian officers, university publications from Brussels and whatever books he could lay hands on. It was a habit he retained all his life. He knew tracts of the Bible off by heart. Later, his regular favourites were to give a clear indication of the sense of personal destiny that had developed: President Charles de Gaulle, Winston Churchill and Niccolò Machiavelli, author of *The Prince*, that autocrat's handbook.

He took and passed an accountancy course and began to dabble in journalism, something he had already practised at school, where he ran the class journal. And he got married. Marie Antoinette, an appropriate name for the wife of a future African monarch, was only fourteen at the time, but in traditional Congolese society this was not considered precocious. Still smarting from his schoolroom clashes with the priests, Mobutu chose not to wed in church. His contribution to the festivities—a crate of beer—betrayed the modesty of his income at the time.

Photos taken during those years show a gawky Mobutu, all legs, ears and glasses, wearing the colonial shorts more reminiscent of a scout outfit than a serious army uniform. Marie Antoinette, looking the teenager she still was, smiles shyly by his side. Utterly loyal, she was nonetheless a feisty woman, who never let her husband's growing importance cow her into silence. 'You'd be talking to him and she would come in and chew him up one side and down the other,' said Devlin. 'She was not impressed by His Eminence, and he would immediately switch into Ngbandi with her because he knew I could understand Lingala or French.'

A Belgian colonial had started up a new Congolese magazine, *Actualités Africaines,* and was looking for contributors. Because Mobutu, as a member of the armed forces, was not allowed to express political opinions, he wrote his pieces on contemporary politics under a pseudonym. Given the choice between extending his army contract and getting more seriously involved in journalism, he chose the latter. Although initial duties involved talent-spotting

Congolese beauties to fill space for an editor nervous of polemics, Mobutu was soon writing about more topical events, scouring town on his motor scooter to collect information. The world was opening up. A 1958 visit to Brussels to cover the Universal Exhibition was a revelation and he arranged a longer stay for journalistic training. By that time he had got to know the young Congolese intellectuals who were challenging Belgium's complacent vision of the future, staging demonstrations, making speeches and being thrown into jail.

One man in particular, Lumumba, became a personal friend. The two men shared many of the same instincts: a belief in a united, strong Congo and resentment of foreign interference. Thanks to his influence Mobutu, who had always protested his political neutrality, was to become a card-carrying member of the National Congolese Movement, the party Lumumba hoped would rise above ethnic loyalties to become a truly national movement.

But even in those early days there are question marks over Mobutu's motives. Congolese youths studying in Brussels were systematically approached by the Belgian secret services with an eye to future cooperation. Several contemporaries say that by the time Mobutu had made his next career step—moving from journalism to act as Lumumba's trusted personal aide, deciding who he saw, scheduling his activities, sitting in for him at economic negotiations in Brussels—he was an informer for Belgian intelligence.

What were the qualities that made so many players in the Congolese game single him out? Some remarked on his quiet good sense, the pragmatism that helped him rein in the excitable Lumumba when he was carried away by his own rhetoric. It accompanied an appetite for hard work: Mobutu was regularly getting up at 5 in the morning and working till 10 p.m. during the crisis years. But the characteristic that, more than any other, eventually decreed that he won control of the country's army was probably the brute courage he attributed to that childhood brush with the leopard.

Bringing the 1960 mutiny to heel involved standing up in front of hundreds of furious, drunk soldiers who had plundered the barracks' weapons stores and quelling them through sheer force of personality. And Mobutu carried out that task, one that civilian politicians understandably balked at, not once but many times. 'I've been in enough wars to know when men are putting it on and when they really are courageous,' said Devlin. 'And Mobutu really was courageous.' Once, he watched Mobutu curb a mutiny by the police force. 'They were hollering and screaming and pointing guns at him and telling him not to come any closer or they'd shoot. He just started talking quietly and calmly until they quietened down, then he walked along taking their guns from them, one by one. Believe me, it was hellish impressive.'

The quality was to be tested repeatedly. The assassination attempt foiled by Devlin's intervention was one of five such bids in the week that followed Mobutu's 'peaceful revolution'. Such was the danger that Mobutu sent his family to Belgium. Marie-Antoinette deposited her offspring and returned in twenty-four hours, refusing to leave her husband's side. 'If they kill him they have to kill me,' she told friends.

What constitutes charm? A presence, a capacity to command attention, an innate conviction of one's own uniqueness, combined, as often as not, with the more manipulative ability of making the interlocutor believe he has one's undivided attention and has gained a certain indefinable something from the encounter. Whatever its components, the quality was innate with Mobutu, but definitely blossomed as growing power swelled his sense of self-worth. In the early 1960s European observers referred to him as the 'doux colonel' (mild-mannered colonel), suggesting a certain diffidence. Nonetheless he was a remarkable enough figure to prompt Francis Monheim, a Belgian journalist covering events, to feel he merited an early hagiography. By the end of his life, whether they loathed or loved him, those who had brushed against Mobutu rarely forgot the experience. All remarked on an extraordinary personal charisma.

'I've never seen a photograph of Mobutu that did him justice, that makes him look at all impressive,' claimed Kim Jaycox, the World Bank's former vice-president for Africa, who met Mobutu many times. 'It's like taking a photograph of a jacaranda tree, you can't capture the actual impact of that colour, of that tree. In photos he looked kind of unintelligent and without lustre. But when you were in his presence discussing anything that was important to him, you suddenly saw this quite extraordinary personality, a kind of glowing personality. No matter what you thought of his behaviour or what he was doing to the country, you could see why he was in charge.'

He had a gift for the grand gesture, a stylish bravado that captured the imagination. Setting off for Shaba to cover the invasions of the 1970s, foreign journalists would occasionally disembark to discover, to their astonishment, that their military plane had been flown by a camouflage-clad president, showing off his pilot's licence.

There were some of the personal quirks that can count for much when it comes to political networking and pressing the flesh, whether in a democracy or a one-party state. He had a superb memory and on the basis of the briefest of meetings would be able, re-encountering his interlocutor many years later, to recall name, profession and tribal affiliation. 'It was phenomenal,' remembers Honoré Ngbanda, who as presidential aide for many years was responsible for briefing Mobutu for his meetings. 'Whether it was a visual memory or a memory for dates, he could remember things that had happened 10 years ago: the date, the day and time. His memory was elephantine.'

Mobutu had another of the characteristics of the manipulative charmer: he could be all things to all men, holding up a mirror to his interlocutors that reflected back their wishes, convincing each that he perfectly understood their predicament and was on their side. 'He could treat people with kid gloves or he could treat them with a steel fist,' remembered a former prime minister who saw more of the fist than the glove. 'It was different for everyone. He was very clever at tailoring the response to the individual.'

Not for him the rigid stances that had doomed Lumumba. He would dither for days, leaving his collaborators in a state of nervous ambiguity, often uncertain over what instructions had actually been issued. This was the negative side of his adaptability. But while colleagues tried to second-guess his wishes, he would be assessing the mood of the day, ready to change direction with all the panache of a born actor. 'He was very good at putting on a show,' acknowledged a contemporary. 'He could be absolutely furious and two minutes later, when he saw it wasn't the right thing to do, he'd change completely.'

And finally, there was the humour: sardonic, worldly wise, it deepened as the years turned against him until, listening to Mobutu fielding questions about human rights and corruption at a hostile press conference in Biarritz, it was difficult not to feel a certain grudging admiration for the impeccable politeness, the fake innocence, the ironic demeanour that all broadcast one defiant message: I know your game and I am far too old and wily a fox to be caught out.

This was the man who seized control of Congo in September 1960. He was to prove as good as his word, swiftly handing power to a group of 'general commissioners'—a collection of the country's few university graduates—who were supposed to run the country while the politicians took stock of the problems confronting Congo. With four separate governments in existence—one in the eastern city of Stanleyville, loyal to the ousted Lumumba; one in Katanga under Moise Tshombe, supported by the Belgians; one in Kasai under Albert Kalonji; and one in Leopoldville under President Kasavubu—national partition was now a reality rather than a threat. But the disappearance of probably the key player in this game was about to alter the situation.

In the space of a couple of months, Lumumba had managed to outrage the Belgians by insulting their king, appal the West with his flirtation with Moscow and alienate the United Nations. He had also frightened former colleagues by hatching a series of cack-handed assassination plots against his Congolese rivals. With Mobutu in charge, Lumumba was now in detention, but his Napoleon-like ability to whip up the crowds and convert waverers to his cause—even at times his own jailers—meant he remained a dangerous loose cannon.

In August of that year, the CIA director himself had told Devlin that Lumumba's removal was an 'urgent and prime objective', an instruction that presumably could have covered anything from encouraging Lumumba's rivals to topple him by legal means to funding a *coup*. Now Washington moved to direct action. Shortly after Mobutu's takeover, Devlin was advised by headquarters that 'Joe from Paris' would be coming to Leopoldville on an urgent mission. 'I was told I'd recognise him, and I did. He was waiting at a café across from the embassy and he walked me to my car and we went to a quiet place where we could talk.' The man was a top CIA scientist and he had come to Kinshasa with a poison for Lumumba. Devlin, he said, was to arrange for it to be slipped into the prime minister's food, or his toothpaste. The poison was cleverly designed to produce one of the diseases endemic to central Africa so that Lumumba's death would look like an unfortunate accident. 'Jesus Christ, isn't this unusual?' was Devlin's astonished reply. Joe from Paris acknowledged that it was, but said authorisation came from President Eisenhower himself.

It was a job the usually conscientious Devlin somehow never got around to performing. He insisted, and has testified before a US Senate committee hearing, that while he held no moral objections to the principle of political assassination when demanded by circumstances, the killing of Lumumba was never a step he personally considered necessary or intended to carry out. 'If I had had Hitler in my sights in 1941 and I'd pulled the trigger, maybe 20 or 30 million people would be alive today. But I just never felt it was justified with Lumumba. I was hoping the Congolese would settle it amongst themselves, one way or another.'

No hint of his doubts, it must be said, appeared in his cables, raising the question of whether Devlin really was as reluctant as he makes out or just sanitised his account with the passage of time. Although he had access to Lumumba's entourage, Devlin stalled. The months passed, with the CIA considering first one assassination scenario and then another. Devlin eventually disposed of the poison by pouring it into the Congo river. 'I had the damn stuff in my drawer and I wanted to get rid of it.'

During that time, Lumumba was demonstrating just what a threat he remained by dramatically escaping from detention. Recaptured, he was transferred to a military base, only to be briefly freed when the soldiers mutinied. Finally, both the Americans and Belgians were provided with the let-out they had been hoping and pushing for all along. Exasperated by Lumumba's Houdini-like qualities, Kasavubu and the commissioners dispatched the prime minister to his arch-enemies in the south. On 17 January 1961, Lumumba and two collaborators were flown to Katanga. During the flight down, Luba soldiers avenged themselves for the massacre of their tribesmen by Leopoldville's army the year before, beating their prisoners so brutally the horrified Belgian air crew closed themselves in the cockpit to drown out the noise. Approaching Elizabethville, the pilot radioed the control tower to announce 'I have three precious packages aboard'. On arrival, they were taken away to be killed, almost certainly shot dead in front of Katanga's top officials and their Belgian collaborators.

The Congolese secessionists had done the CIA's dirty work for them. Devlin insisted he was not aware of Lumumba's departure until it was too late. As for Mobutu, no one has ever been able to prove his involvement in the murders. The adoring account written by the Belgian Francis Monheim claims, unconvincingly, that Mobutu was never even told of the decision to move the prisoner. But it is almost impossible to believe that the head of the army, the man who held the real power in the country, could have been kept unawares of such a key development. 'I can't believe he wasn't involved,' confessed Devlin. 'But it was just one of those questions you didn't ask at the time.'

Whoever actually pulled the trigger, in the eyes of Lumumbists and many other Zaireans, Mobutu always bore moral responsibility for Lumumba's murderer, with the Western powers playing the part of Iago, whispering their instructions from behind the scenes. Mobutu's decision later to erect a monument to the country's first prime minister was regarded as an act of extraordinary cynicism, Orwellian in its apparent intention of rewriting history. Certainly, the story of Lumumba and Mobutu follows the pattern of one of the great parables of mankind: the loving brothers, the best friends who end up trying to destroy each other, their former intimacy ironically rendering them more ruthless, more implacable in their hatred than any two strangers could ever be. It is the story of Romulus and Remus, Cain and Abel, Macbeth and Banquo. There was even a moment during those years when a dishevelled Lumumba, thwarted by Mobutu in a bid to address the army, turned on his former friend and in a quiet, sad voice said: 'Is it you, Joseph, saying that?' And Mobutu replied: 'Yes, it is me. I have had enough.' It was the 1960s equivalent of 'Et tu, Brute?' from the dying lips of the betrayed Julius Caesar.

Lumumba had certainly started off being the dominant member of the partnership, more famous, more charismatic, more politically sophisticated and far more idealistic. But he had lacked pragmatism, and that was Mobutu's forte.

The whereabouts of Lumumba's body have never been identified. It was probably hacked into pieces, the head dissolved in a vat of sulphuric acid by a Belgian clean-up team sent to remove all traces of the assassinations. But another, even more fanciful story has done the rounds: that Mobutu's collaborators, terrified that Lumumba's spirit would live on after his death, asked a witch-doctor how to destroy his supernatural powers. On his instructions they divided up the body, hired a low-flying C130, and flew along the borders of their huge country, scattering the pieces. This was the only way, the marabout had said, to prevent Lumumba's spirit reassembling and returning to challenge his former friend.

Lumumba's death removed a man who, while alive, would always represent a challenge to those in power. But it did not end Congo's political turmoil. Sucked into the government's attempts to re-establish territorial integrity, the UN was to find itself embroiled in pitched battles with a mercenary force recruited by the Katangans and to lose its secretary-general when Dag Hammarskjöld was killed in a plane crash flying to yet another round of negotiations.

Once the Katanga and Kasai secessions had been brought to heel, the government was confronted by a new set of anti-Western, Marxist uprisings in the east. In one of these a young rebel called Laurent Kabila, whose womanising ways and heavy drinking exasperated Che Guevara, the Argentine revolutionary who had set off to help his African brothers, played a role before fading from view. Western audiences were more preoccupied by the horrors that occurred in Stanleyville, where a white mercenary force and Belgian paratroopers were unable to prevent the slaughter of 200 Europeans held hostage by the rebels. Never again would Western leaders and the public at large look at Africa with the same cheerful optimism of the postwar days.

During all this time Mobutu, as head of the armed forces, was watching and waiting, a quiet presence behind the succession of weak and divided civilian governments. By October 1965, another political impasse had developed, with Kasavubu sacking Tshombe, the rebel-turned-prime minister, and elections looming. By this stage, Mobutu had become a regular visitor at the Devlin household. The two men had got to know each other's families, with Mobutu taking a particular shine to the CIA station chief's young daughter, who liked to steal his cap and swagger stick and march up and down with them. However, what happened in November, Devlin maintained, was not the result of any advice on his part. 'The US position and British position was that they did not want a *coup*, they wanted Kasavubu as president and Tshombe as prime minister. I told Mobutu that, and he smiled and said: "A Johnson–Goldwater ticket you mean?" (a Democrat–Republican combination that would have united the US's two main parties) I said "Yes", and he said "Fine". The next thing I knew, I was woken at five in the morning with the word he had just pulled a *coup*.'

This time, despite promises to hand power back to the civilians in five years' time, Mobutu was not going to modestly bow out. He was there to stay—for thirty-two years. With hindsight and the knowledge of what was to come, it is too easy today to forget that the second Mobutu takeover was welcomed, not only by foreign allies desperate to see a safe, pro-Western hand on the tiller. Disillusioned by five long years of wrangling and war, the Congolese had lost all faith in the efficacy of armed struggle. They ached for a stability the civilian politicians seemed incapable of achieving. Mobutu's delivery of just that quality was to render him massively popular for years to come.

Devlin left Congo in 1967 for Laos, where he was to win the Distinguished Intelligence Medal he wore with quiet pride for a particularly risky battlefield operation. He was not to rebase in Kinshasa until 1974, by which time he had left the CIA, although he deemed it hardly worth his while, given the high public profile he had inadvertently acquired in Congo, to even attempt to cover his tracks. 'You can retire from the agency under cover if you want to. But I told them that in my case it would be a bit like a whore who'd worked the same block for twenty years coming back as a nun.' He was taken on by Maurice Templesman, the secretive diamond dealer. But although he brandished a letter from the CIA's director stating he was no longer with the agency, Mobutu still invited him around. 'I wanted it to be clear I was no longer in the business. But he liked to use me as a sounding board and perhaps, sometimes, to carry a message back to Washington.'

Devlin found a changed man. Austere army barracks had been replaced by the comfort of the presidential villa on Mount Ngaliema, the hill overlooking the Livingstone falls, or the burgeoning palace at Gbadolite. The 5 a.m. starts were still being observed. But Mobutu, exposed to only beer before, had learned to enjoy his drink. And in a sign of the luxury he had come to regard as his due, his choice was pink champagne, the Hollywood movie star's tipple. 'He was already round the bend, more paranoid, more convinced of his own infallibility. He was surrounded by yes-men who were constantly telling him how wonderful, how brilliant, how marvellous he was, what an extraordinary mind he had. All I could think of were the stories I'd read about the court of Henry VIII or Louis XIV.'

Like many men of his era, Devlin felt he had no apologies to make for past policy decisions. It was too easy, he insisted, for a new generation to forget the very real imperatives of the day. 'You're too young to remember much about the Cold War. But it was a real war and Mobutu played a rather key role in blocking Khrushchev. He was right for Congo at that time.'

Yet that did not blind him to what the 'doux colonel' had become by the 1970s: 'Lord Acton had it exactly. "Power corrupts and absolute power corrupts absolutely." '

CHAPTER FOUR

Dizzy worms

'If you find excrement somewhere in the village, the chief was the one who put it there.'

Bas-Congo proverb

'There are no opponents in Zaire, because the notion of opposition has no place in our mental universe. In fact, there are no political problems in Zaire.'

—Mobutu Sese Seko

For more than a decade, a fanatical rebel movement called the Lord's Resistance Army (LRA) has been operating in Uganda's impoverished north. Led by Joseph Kony, a crazed former choirboy who says he wants to rule the country according to the Bible's Ten Commandments, it is feared and hated by local villagers for the atrocities it commits.

Despite its unpopularity, the LRA has successfully challenged the Ugandan government, running rings around army units sent to quell it, and halting development in the region. Its success can partly be attributed to the peculiarly unpleasant technique used to recruit new members. An LRA unit will target a school and force the pupils to march off with its fighters. The girls are then raped and taken as wives by LRA commanders. The boys are given bayonets and ordered to kill a fellow pupil, a playground friend, who may have shown signs of straggling. Once the crime has been committed, the children have blood on their hands. The disgust they feel for themselves, their terror of the vengeance the village elders would mete out if they knew the truth, prevent them from returning to civilian life. It is guilt by association, and it is a terribly effective method for extracting loyalty from even the most reluctant.

There was something of the LRA technique, the methodology of the vampire initiating his latest victim to the secret world of the undead, about the way Mobutu set about the task of consolidating his position once the mutinous army had been brought under control. In the years that followed he was to adopt a variety of techniques to shore up his rule, ranging from terror, to divide and rule, to sheer demagoguery. But his appeal to one of the most powerful of all human instincts—greed—was to prove by far the most effective means of co-opting a generation.

As a member of one of the smaller of Zaire's 250 or so tribes, he knew he could not count on the automatic support of any sizeable ethnic community. His country was intimidatingly large. If the secessions had been brought under control, local governors remained unruly, neighbouring powers itching for a chunk of the land he could barely police. The likelihood of a *coup* leader managing to stay at the helm beyond the year seemed slim indeed.

Early on in his regime there were public hangings of suspected *coup* plotters, with the public encouraged to attend the gruesome open-air spectacle. Pierre Mulele, the rebel leader who had challenged central rule in the east, was lured back from exile with an amnesty promise, then tortured to death by soldiers. His eyes were pulled from their sockets, his genitals ripped off, his limbs amputated one by one as he slowly expired. What remained was dumped in the river.

But these were the crude, traditional methods a new leader used to show who was boss. By the mid-1970s, Mobutu had grown more subtle. Why kill your enemies, after all, when, with a bit of financial encouragement, they would willingly sell their souls? It was an approach that elided smoothly into his own conception of his role as a tribal chief who, like a gangster boss, must be able to prove his value to the community in concrete terms. 'If you go to see the head of the village you never come back empty-handed,' Mobutu would say, handing over funds for a school here, a hospital there. 'Those who come to see me must always go away with something.'

He had attended the round-table talks staged to settle outstanding economic disputes between Congo and Belgium before independence. Looking back, he became convinced the naive Congolese delegation had been diddled by the wily white negotiators. But the experience had left him with a keen appreciation of the extent of his country's assets he could put to good use. The pie waiting to be divided up was enormous. And it could be made yet bigger if a sensitive issue was addressed: ten years after nominal autonomy, 75 per cent of the country's economy was still in foreign hands. Psychologically, culturally and economically, Congo was still under colonial sway. It was time for a redistribution of wealth that would right past wrongs and simultaneously create an elite who would owe Mobutu everything and be suitably grateful.

Unfortunately it was here that one of his great personal failings was to be exposed. A self-made man, Mobutu was bright, quick to learn. Like many an African president he had risen to the top by dint of sheer toughness and a cunning understanding of human behaviour. His was the wiliness of the street operator, not the analytical intelligence of the academic. He had never completed his formal education or gone to university. His mental landscape was a jumble of half-learned lessons, gut convictions and practical wisdom, lacking structure and discipline. Though he was ready to fill certain gaps by reading up on political strategy and military tactics, the same did not apply to another, equally important field—economics.

Official after official attests to the fact that when the subject of economics came up, Mobutu's attention would wander, his eyes glaze over. He knew how much his country was worth, but he had no idea of the processes required to realise that value. As Oscar Wilde might have said, he knew the price of everything while understanding the value of nothing. 'He had absolutely no interest in economics,' acknowledged Jose Endundu, who was one of the country's leading businessmen during the Mobutu years. 'He didn't understand that without a sound economy there is no politics. He couldn't see the link. If you tried talking economics he'd immediately change the subject. If you brought him an economic document he'd give it to an aide and say "put it on my pillow, I'll read it later", but you could be sure he would never look at it.'

Mobutu was out of his depth, but was not prepared to admit it. 'If you know nothing, you let others manage,' said a former prime minister, drawing an analogy with France's great Sun King. 'Jean-Baptiste

Colbert, the economic adviser to Louis XIV, would say: "Sire, you handle the politics, and I will manage the economy." But Mobutu would not do that.'

This weakness was compounded by the changing nature of his coterie. By the 1970s, those who visited him noted that advisers willing to offer cogent criticisms had disappeared, replaced by sycophants ready to say whatever they thought the great man wanted to hear. When it came to politics or the army, Mobutu at this stage of his life was too astute to be misled by such yes-men. But in the field of economics, where his boredom threshold was low, he would grasp at the miracle cures being offered by members of his entourage. Long-term consequences went unexamined, knock-on effects ignored in the rush to tie things up. 'He liked easy options. If someone came to him and offered him what looked like a nice, easy solution he'd seize it. And like many people with limited education, he wouldn't know when to let go,' said Devlin. 'He was a political genius, but an economic spastic.'

In the leader of a military putsch, such a failing was trivial. In a president who was to spend three decades in power, it proved devastating.

As with most autocrats, Mobutu's personal charisma went hand-in-hand with an instinctive feel for the masses. It was an understanding he carefully nurtured in the first fifteen years of his rule, travelling the country constantly in his determination to fuse the fractious provinces into one nation. 'His party piece was to call some regional governor and announce he would be flying into his district at noon. It was his way of keeping them on their toes,' recalled former US ambassador Daniel Simpson, who did a total of three tours of the country.

There were frequent rallies in sports stadia and halls, at which 'Papa' would talk to his children. As every member of Congolese society automatically belonged to the Popular Revolutionary Movement (MPR), the party he had founded, attendance was recommended. The public came expecting entertainment, and Mobutu would oblige. Like a pantomime performer drawing the crowd into his oh-no-you're-not, oh-yes-you-are routine, he would warm his audience up with a question-and-answer session they came to anticipate:

Mobutu shouts: Nye, nye? (Can you be silent?).
Crowd roars: Nye (We are silent).
Mobutu: Na Loba? (Can I speak?).
Crowd: Loba (Speak).
Mobutu: Na Sopa? (Can I speak frankly?).
Crowd: Sopa (Speak frankly).
Mobutu: Na Panza? (Can I speak openly?).
Crowd: Panza (Speak openly).

Then would follow a speech in Lingala, the language which, unlike the French mastered by only an educated elite, was accessible to the common man. It would be full of puns, wordplay and wisecracks. Mobutu would get the crowd giggling, cheering and laughing. As often as not, there would be a public putdown for an unpopular aide or minister, sometimes a sacking. It was Mobutu's way of assessing the national mood and lancing the boil of public discontent before it turned septic.

'He was a speaker of genius,' said a Congolese journalist who was a student at the time. 'I would go unwillingly, because I didn't really approve of Mobutu. But as soon as he began speaking, we would be swept away. We'd stand in the sun for hours, but the time would slip by without you noticing. If you study those speeches now, in the cold light of day, you can see there was almost nothing in them, they were full of inconsistencies, gossip and tittle-tattle. But he knew just how to speak to the people. He would tell us nonsense and we would believe him.'

It was that demagogic talent Mobutu was to exploit in the early 1970s as he launched the most intellectually ambitious project of his career. Five years after his military *coup*, he was increasingly aware that he needed to give his one-party regime an ideological lick of paint. The secession attempts of the 1960s had shown how prone to fragmentation the country was. Some binding, unifying philosophy was needed. There was also an element of one-upmanship to what followed. The post-independence era had seen a flourishing of African sensibility. From Ghana, Kwame Nkrumah's vision of black consciousness and pan-Africanism had spread across the continent. In Tanzania, Julius Nyerere had launched his doctrine of socialism and self-reliance. In Senegal Leopold Senghor had preached the proud doctrine of 'negritude'. As the leader of the third largest African nation, with the awareness of all those incredible mineral riches at the back of his mind, Mobutu felt greatness was rightfully his. His country would play a leading role in the non-aligned movement, become the West's preferred African interlocutor and act as a catalyst for change on the continent and beyond, he decided. His head ringing with the praises of his entourage, the self-taught former army sergeant launched himself into a campaign he was signally ill-equipped to see through. Drawing on concepts of mass mobilisation, strong leadership and revolutionary 'militancy' picked up during visits staged to a friendly China and South Korea, he set out to take the intellectual lead in Africa.

At its best, 'authenticity', as the movement born in 1971 was called, was an admirable attempt to recover a sense of African identity and pride crushed by the colonial experience. It was remarkable, in a way, that it had taken so long for a new country to feel the need for a fresh image. The country must modernise, Mobutu told his public rallies, but it would do so in a framework of ancestral spiritual values, not by aping Western materialism. 'Authenticity is the realisation by the Zairean people that it must return to its origins, seek out the values of its ancestors, to discover those which contribute to its harmonious and natural development,' Mobutu told the United Nations. 'It is the refusal to blindly embrace imported ideologies. It is, in short, the affirmation of mankind, in its place, as it is, with its mental and social structures.'

Congo was rebaptised Zaire, and the national currency and main waterway similarly renamed: one trademark embraced three key concepts. The Christian names left by the European missionaries were abandoned and African names revived, with Mobutu setting a personal, grandiose example. Roads and squares named after Belgian notables were rebaptised after key events in the struggle for independence and the national anthem and flag were changed. This was the moment when the statues of Stanley, King Leopold and King Baudoin were toppled.

'Madame' and 'Monsieur' were replaced by 'Citoyen' and 'Citoyenne', in an echo of the French Revolution. Instead of the European suit, men were to don a high-collared jacket of Mobutu's invention. Dubbed the abacost (from 'à bas le costume'—'down with the jacket'), and usually modelled in dark brown or navy blue wool, this was no better adapted to the African climate, but it was different. Returning Zaireans would whip off their ties in the plane for fear of having them snipped in two at customs. As for women, the provocative miniskirts of the 1970s were replaced by more dignified African pagnes, or wraps, and wigs shunned in favour of 'natural' coiffure.

Such changes hardly amounted to a coherent philosophy. But then, there probably never was one within the Mobutu brain, which stalled over even the basic question of the MPR's political orientation, defined as being 'neither left nor right, nor even the centre'. 'If he had focalised and crystallised his thought by writing it down, there were rich ideas there waiting to be developed,' insisted Honoré Ngbanda, who later became one of Mobutu's closest aides. 'It was a fundamental philosophical notion. But unfortunately, whether it was at the level of the MPR's central committee, the government or his own collaborators, there was no one who could take the idea and give it a conceptual form.'

However quaint its manifestations, authenticity did make its mark. For many Congolese today, it is the one Mobutu gift for which they remain grateful, leaving them with a sense of uniqueness, the awareness that they were not Kasaians, Shabaians or natives of Bas-Congo, but citizens of one vast central African nation with its own, very distinct identity.

Even sceptics were won round. 'At first all that business with the abacosts just seemed ridiculous to me,' said Devlin. 'Then I caught on. It meant "we are all one, I am a great leader and I have the respect of the world." It did make Zaireans feel they were something special.' So special, indeed, that the organisers of the world heavyweight boxing match between Muhammad Ali and George Foreman chose Kinshasa as venue for a 'Rumble in the Jungle' in 1974 that was as much a celebration of budding black pride as a sporting event.

Unfortunately, the movement could all too easily be pushed in another direction, in a climate where Mobutu's aides were falling over themselves to demonstrate their loyalty and commitment. Swiftly, authenticity elided into Mobutuism—never really defined—and an extravagant personality cult. The Guide, the Helmsman, Father of the Nation, Founding President, the official media called him. Thanks to the efforts of Sakombi Inongo, his public relations maestro, Mobutu's face, with its carp-lips and heavy glasses, was on the cover of almost every newspaper. The daily television news broadcast began with an image of his features, emerging God-like from scudding clouds and his arrival was met with dancing and singing. Officials went so far as to compare Mobutu to the messiah, with the MPR as his church and party cadres as the disciples.

But Mobutu was no Mao and the infinitely pragmatic Congolese were not in the same mould as the Chinese. If they had learned all about subjugation and passive resentment under Leopold, blind reverence, the wholesale swallowing of propaganda, was quite another matter. Mobutuism died a quiet death as the personality cult was met with a sceptical shrugging of shoulders and soft amusement.

The same could not be said of the campaign that acted as the economic corollary of authenticity: Zaireanisation. In 1973 Mobutu decreed that foreign-owned farms, plantations, commercial enterprises—mostly in the hands of Portuguese, Greek, Italian and Pakistani traders—should be turned over 'to sons of the country'. This was followed by radicalisation, in which the largely Belgian-controlled industrial sector was confiscated. The president may have envisaged some Chinese-style return to the land which would allow an alienated urban class to rediscover its ancestral roots and spell an end to rural stagnation. In theory, departing foreigners were to be compensated and the performance of new owners would be carefully monitored. In practice—a result, perhaps, of that presidential lack of attention when it came to matters financial—no guidelines were drawn up to specify who got what. The result was an obscene scramble for freebies by the burgeoning Zairean elite. Thousands of businesses, totalling around $1 billion in value, were divided between top officials in the most comprehensive nationalisation seen in Africa.

Mobutu, of course, did best out of the share-out, seizing fourteen plantations which were merged into a conglomerate that employed 25,000 people, making it the third largest employer in the country, responsible for one fourth of Zairean cocoa and rubber production. Members of his Ngbandi tribe were the next to benefit, their plum positions in the newly nationalised companies and prime enterprises making up for all those jibes about rural backwardness. But Mobutu was careful to ensure all the major ethnic groups whose support he needed profited. The social class known as the 'Grosses Legumes' (Big Vegetables)—a term used by ordinary Zaireans with a mixture of resentment and awe—was born.

One Congolese woman, who was seventeen at the time, remembers having a surreal conversation with the father of a schoolfriend, who was helping to distribute the seized businesses: ' "Would you like a shop?" he asked me. When I said "no", he insisted. "Go on, take one, I'll give you the staff." When I still said I wasn't interested, he said: "Look, if you don't want it yourself, you can always give it to your mother." '

It was an approach to economic management so naively simple in its conception, so shoddily applied, that it was doomed to failure. Many of the new owners, busy forging their careers in Kinshasa, found they had neither the skills nor the interest to run their new toys. 'They were giving cattle herds to people who couldn't sex a bull,' scoffed Endundu, the businessman. 'It was a disaster.' Others never even bothered to try and make a go of it, convinced the manna from heaven would keep dropping painlessly from on high. As expatriate managers headed for the airports, the new owners pocketed savings, sold herds, dumped equipment on local markets and ripped up bushes.

The proceeds were spent on luxury items, with imports of Mercedes-Benz hitting an African record one year after Zaireanisation. Ordinary Zaireans, supposed beneficiaries of the process, watched in shock as businesses closed, prices rose, jobs were doled out by new bosses on purely nepotistic lines, and shelves emptied. After a few weeks, the Big Vegetables could be heard asking when the Portuguese and Greek businessmen who had left the country would be coming back to restock the warehouses. But in the West, outraged parent companies had halted supplies and frozen credit to their former subsidiaries.

Mobutu could not have picked a worse moment to deliver such a shock to his economy. Intoxicated by his sense of national greatness, he had already borrowed heavily for a series of pharaonic industrial projects, including the building of the dam at Inga and the installation of a transmission line linking it to the southern mining area. In 1974, the world price for copper dropped by nearly two thirds as the oil shock plunged the world into recession.

Until Zaireanisation, the economy had grown by an average 7 per cent a year. Look at a graph of just about any indicator and there, in 1974, is the sharp peak, followed by a long, slow, unstoppable swoop that continues to this day. 'Mobutu survived an extraordinary decline,' said Jerome Chevallier, a former World Bank resident representative. 'But he killed that economy.'

When the disastrous implications of what he had done began to sink in, Mobutu tried to reverse the damage. Majority shares were offered to former owners and some compensation finally paid. But no entrepreneur, large or small, would ever invest in the country with a tranquil heart again. Those who had put their faith in Zaire had been looted on the president's orders. In future, businessmen who ventured there would demand exceptionally high rates of return and quick profits to justify the risk they were running, and would be careful to repatriate their proceeds rather than reinvest.

The blow to Mobutu's notions of grandeur was multi-pronged. For the economic crunch came as Africa's leaders rejected authenticity as a model, contemptuous of Mobutu's attempts to portray himself as a great thinker. Hurt by the hostile international reception, recognising that he had badly overstretched himself, Mobutu was never subsequently to attempt to give his rule ideological content. From then on, he would be the ultimate pragmatist, concerned only with what was necessary to keep him in power and allow him to make money.

For his population, Zaireanisation's impact was to extend far beyond the immediate commercial crisis. The belief that something could be had for nothing, the looter's smash-and-grab mentality, had been endorsed at the very highest level of society. Mobutu and his ministers had plundered mercilessly, and no one had ever been punished. The president himself—in a slip made during a speech transmitted live on television—even appeared to articulate the new philosophy, telling employees: 'Go ahead and steal, as long as you don't take too much.' The lesson was not lost on those lower down in the hierarchy, in far greater need of extra cash than their superiors.

Before Zaireanisation, corruption, while a problem, had seemed to observers on a par with that witnessed in many other emerging African states. But in the generalised climate of impunity created by this botched economic experiment, sleaze—whether practised by the lowly bread-seller or the Mercedes-driving Big Vegetable—was about to become the most striking characteristic of Zairean society.

Mobutu's appropriation of one of Zaire's largest plantations was the public signal that a new system of rule, glimpsed elsewhere in the world but never, perhaps, brought to such a level of purity, was in the process of being established in Africa: a kleptocracy.

At the top of the heap sat the chief pilferer. Nobody knew for sure the size of his bank accounts, the range of his company holdings and investment funds. Deeds were made out to front companies, business associates and family members in order to cover his tracks.

But it was easy enough to monitor the gradual build-up of a portfolio of desirable residences at home and abroad.

In Zaire itself, the palace complex at Gbadolite gradually blossomed into life like one of those lush tropical flowers which virtually poison the air around, so potent is their scent. In each major town, a villa lay ready for the president's use. Kinshasa, of course, also boasted a choice of residences, including a hillside mansion whose grounds served as a private zoo. But Mobutu came to prefer the pagoda in the Chinese village built at Nsele, east of the capital, or his luxury cruiser *Kamanyola*, furnished with oyster-shaped settees in pink silk.

Some presents Mobutu made himself were simply too ostentatious to remain secret for long. Most notorious was the $5.2 million Villa del Mar in Roquebrune Cap Martin, not far from King Leopold's former French Riviera estate. The story goes that when buying this neoclassical property, the president agreed a price, then as an afterthought enquired whether it would be in dollars or Belgian francs—the 39-fold difference in value held no meaning for a man of such careless wealth.

On a similar lavish scale were Les Miguettes, a converted farmhouse in the Swiss village of Savigny and the $2.3 million Casa Agricola Solear estate in Portugal's Algarve, blessed with 800 hectares of land, a 14,000-bottle cellar and 12 bedrooms. There was also a vast apartment on Paris's Avenue Foch, conveniently close to the furrier who made his trademark leopardskin hats and the fashion designers patronised by his family. From Cape Town to Madrid, Marbella to Marrakesh, Abidjan to Dakar were scattered a string of farms, villas and hotels.

However, the bulk of his real estate network was located in Brussels. The turreted Château Fond'Roy was just one of at least nine buildings scattered across the upmarket districts of Uccle and Rhode St Genèse, a sign that this most nationalist of African presidents was ready to forgive colonial wrongdoings when it came to finding a safe investment for his money. It was an impressive collection for a man who in 1959 claimed to have just $6 to his name.

But no one could accuse Mobutu of hogging it all to himself. In playing to the hilt the role of high-rolling African Big Man, he was simultaneously setting an example for the Big Vegetables to copy. What he wanted was a quiescent, loyal political class and Zaireanisation was only the bluntest of methods of drawing as many potential rivals as possible into the establishment. Under Mobutu, indeed, the entire state administration came to resemble one of those conveyor belts that whizzes an array of gifts past contestants in a television game show, with even disappointed also-rans entitled to generous consolation prizes.

Via the ruling MPR, Mobutu had at least 400 attractive posts, ranging from positions in the party's central committee, executive council, regional administration and state enterprises, to offer as sweeteners. On top of that, an apparently endless succession of cabinet reshuffles prevented ministers from becoming overconfident while allowing a greater number of favourites their turn at the national trough. Between 1965 and 1990, when the one-party system came to an end, Zaire saw fifty-one government teams come and go, an average of two a year. Each contained an average of forty ministers and deputy ministers.

The reshuffles were of no significance when it came to determining Zairean policy either at home or abroad, they simply constituted a way of spreading the largesse around. As one diplomat put it, when asked by a journalist for his assessment of the implications of yet another new government line-up: 'What do you get when you shake up a can of worms? Dizzy worms.'

And the worms were kept very dizzy indeed, their heads reeling with the profits to be made during a tenure they knew was fated to be all too brief. Every official assignment abroad, in itself covered by ridiculously high per diems, was treated like a Christmas shopping trip. Chester Crocker, the former US assistant secretary of state for Africa, would marvel every time he went to the airport to see Mobutu and his aides off. 'The DC10 would barely be able to take off, its belly was so full of stereos and microwaves.'

On ministerial fittings alone, rich pickings were to be had. Each minister had the right to two cars—usually a black Mercedes, Peugeot 605 or all-terrain jeep—while his deputy was entitled to one. Somehow, the cars and furniture were never to be found when the new incumbent moved into his predecessor's strangely depleted offices. Small-scale tribal chiefs themselves, each minister knew he had to live up to the expectations of his constituency, which wanted to see its local hero's worldly success on display. Those favoured knew their duty was to spread such benefits around the extended family. Every new minister, every new head of a state enterprise, would immediately set about distributing jobs to members of his or her ethnic group, appointments that would never be rescinded, merely added to, when a new administration came in.

No wonder that by the 1990s, Zaire had more than 600,000 names on its civil service payroll, notionally responsible for tasks the World Bank estimated could be carried out by a mere 50,000. A perfect example of overmanning was the central bank, which by the 1990s had 3,000 people to shuffle paperwork, more than the 2,000 employed in the country's entire private banking sector.

On top of the flashy cars were the over-generous travelling allowances, the loans from banks that were never—or only partially—repaid, the houses that belonged to the state and were quietly appropriated, the contracts allocated to companies set up by the very officials in charge of ministerial budgets. Mimicking their leader, the Big Vegetables also bought mansions in Brussels and Paris, opened their own Swiss bank accounts, stocked their own cellars with pink champagne. Even those who lambasted Mobutu in public could not resist. For many Zaireans, Cleophas Kamitatu, a vocal critic of the president, set new standards in barefaced cheek when, while serving as ambassador to Japan, he took advantage of high property prices and sold off the Zairean embassy in Tokyo. Kamitatu claimed he made the sale to cover the salaries of unpaid diplomats, but few believed him.

Mobutu turned a blind eye to the accelerating graft. For once the members of an emerging elite had ceded to temptation, once their dirty secrets were logged in the intelligence service's files and stored in his gargantuan memory, they were effectively neutered. In any case, with such profits to be made, with every ambitious graduate convinced he stood a good chance of a ministerial position or chairmanship of a huge state company at worst, the premiership itself at best, what advantage was to be gained by kicking against the system?

The society's brightest and best were sucked into his ambit. Delve into the personal history of almost any Zairean player of significance and you will discover that no matter how talented, how insightful, how articulate he may be about the ills of his society, how apparently determined to correct those faults, he at one stage or another was on the Mobutu payroll. 'I don't blame those guys,' said a diplomat who served in Kinshasa. 'In any society, the most talented people will always go where the money is, whether it's the City, or the media, or Hollywood. In Zaire, the money came from Mobutu. So that was where they went.'

Mobutu favoured collaborators of mixed blood, men like Bisengimana Rema, of Rwandan origin, Kengo Wa Dondo, the son of a Polish magistrate, Seti Yale. Bereft of tribal constituencies, the 'métis' were regarded by many Zaireans as foreigners. Under the constitution, they could never legally aspire to the presidency. They owed everything to Mobutu and he hoped they would remain conscious of the fact. During the early years at least, it was abundantly clear to all supplicants that access to sinecures and favours could be granted by one man alone—the president. This played into another of the great principles on which this admirer of Machiavelli's writings based his domination over an ethnically, linguistically and culturally diverse nation: divide and rule.

Master of the personalised relationship, adept at presenting a different face to each of his interlocutors, Mobutu established intimate links with his subordinates, while working equally hard to ensure they never developed the same rapport with one another. Anything that smacked of a developing cabal was cunningly undermined. Professor Mabi Mulumba made the mistake during his brief stint as premier of inviting the chairmen of four or five state companies to dinner, a gesture judged alarming by a president on the lookout for possible plots. One of his guests was immediately summoned by Mobutu, who told him: 'I see you've been dining with Mabi. Did you know he's asked for you to be sacked?'

On another occasion, Mobutu left a banquet pleading indisposition, an announcement that set the tongues of the generals and ministers assembled around the table wagging on the subject of who would take over if he fell seriously ill. None of these guests, as it happened, as Mobutu was careful to have every word of the conversation relayed to him by a female attendant sent to cater to his needs. 'Later on, one by one, they were all fired,' recalled Mabi. Mobutu loved to be able to embarrass an ambitious apparatchik by bringing up in public some dismissive remark he had made behind the president's back and watching him squirm. The message was clear: his eyes and ears were everywhere. He owned them.

Rumour had it that the president's mastery extended into sexual relations, where Mobutu interpreted his African name—sometimes translated as 'the cock that covers every chicken'—as a licence to help himself to subordinates' wives. So notorious was the president's philandering, political officers at the US embassy were given a daily update of which society hostess was filling the influential post of presidential mistress by their local staff.

At the height of his powers, Mobutu was braced to counter any potential challenge. Working the telephones into the early hours—a one o'clock call from this notorious insomniac was *de rigueur*, call the president's number at 4 a.m. and you could count on an alert Mobutu picking up the receiver himself—he hoovered up gossip, encouraging members of the political elite to inform on each other. As a result, recalls Mabi, the waiting room outside the office where cabinet meetings were held was always curiously silent. 'The first time I went I was astonished, because people weren't talking to each other and they left afterwards without saying a word. They were determined not to provide their colleagues with material that could later be used against them. The system had created its own antibodies. Everyone suspected everyone else.'

Once Mobutu had moved to Gbadolite, group audiences, where Mobutu would have to publicly endorse a viewpoint and a general consensus would emerge, were never held, despite his aides' pleas. Mobutu liked to receive his supplicants separately. This allowed each player to leave his offices believing he had the president's blessing and act accordingly. 'The last person who saw Mobutu was always right,' remembers Pierre Janssen, Mobutu's Belgian son-in-law. 'You would spend all day talking to him, going into details and he would agree with you. You would leave the palace thinking it was all settled and then, if someone came after you, they would win the day.'

Some of this was due to an adaptability that verged on indecisiveness, the lack of confidence of a man who knew, post-Zaireanisation, he would never feature amongst the ranks of Africa's philosopher leaders. But it was also a deliberate attempt to muddy the waters, undermine consensus and thereby prevent the formation of any coherent political movement that might eventually focus on his removal. The subsequent confusion, the furious arguments that followed, would shore up Mobutu's role as ultimate arbiter, underlining the fact that he was the only player who grasped the whole picture. 'He hated those around him to reach agreement,' recalled Honoré Ngbanda. 'He instinctively set people up against each other.'

'I am the king,' he would bluntly tell foreign visitors, drawing a direct line between the absolute control exercised by Belgium's Leopold and his own autocracy.

Pulling strings of jealousy, rivalry and cupidity, Mobutu prevented the emergence of any dauphin who could be embraced as an alternative by Western allies when they began tiring of his rule. It was the old 'who else?' question that was to stymie diplomats in so many African nations. A US State Department official confronted the problem when Bill Clinton assumed the presidency, theoretically bringing a fresh eye to bilateral relations. 'Someone said: "we have to find a Zairean who hasn't been tainted by Mobutu." I burst out laughing and said: "Who?" '

There was no more striking example of the maxim that every man has his price than Nguz Karl i Bond, a politician who made the mistake of appearing to both the Zairean public and the West as a possible presidential successor in the late 1970s. Accused of harbouring designs on the first lady and helping to plot the invasion of Shaba, the pock-marked former prime minister was jailed, condemned to death and tortured so savagely he was said to have been left impotent.

Pardoned, he went into exile, where he tried to unite the gathering opposition movement, denounced Mobutu and his system in a book as 'the incarnation of Zaire's evil', testified against the president before a US Congressional hearing and provided the former IMF executive Erwin Blumenthal with incriminating information about the president's financial misdoings.

Yet by 1985 Nguz had tired of the thankless life of the exile. He returned to Kinshasa to rejoin the MPR fold and Mobutu, who must have revelled in the spectacle of this opposition firebrand defending the very regime he had denounced before a Western audience, named him first foreign minister, and then prime minister—a surrender of personal integrity reported to have cost Mobutu, according to Radio Trottoir, a cool $10 million.

'There was no opposition in Zaire,' agreed Nzanga, Mobutu's son. 'My father used to say "keep your friends close, but your enemies closer still". Leaving people in exile was a danger, they were making a lot of noise. The game was to neutralise their capacity to damage him. So they came back and one by one, I saw all those guys up in Gbadolite. My father would laugh about it. He would say "politics is politics". But he didn't respect any of them.' Nor did the Zairean public, aghast at the crassness of betrayals perpetrated without, it seemed, so much as a flicker of embarrassment.

As the years passed, however, the price of loyalty rose. After Nguz came other high-profile exiles, such as maverick Mungul Diaka, who needed to be silenced. Yet the economic crisis was making it difficult for Mobutu to keep buying popularity. As the collapse of the Berlin Wall dissolved the neat Cold War lines drawn across Africa, the Big Vegetables were beginning to look decidedly itchy.

In April 1990, giving in to popular discontent and growing pressure from the West for reform, Mobutu took an enormous gamble. He resigned his position as head of the MPR and declared Zaire a multiparty state. It was a move many observers blithely assumed spelt the beginning of the end for the Leopard. Etienne Tshisekedi, the mulish former interior minister who headed the outlawed Union of Social and Political Democracy, was seen as probable successor. There was a sudden explosion of the printed media, sensationalist rag-sheets which did not mince their words when it came to ridiculing the president. So confident was the political counsellor at the US embassy that an era was ending, he bet the local CIA station chief Mobutu would not last the year. He was six years out. Drawing on his experience of post-independence anarchy, Mobutu was to become as adept at manipulating democracy as he had been at manipulating single party rule.

As a Sovereign National Conference (CNS) convened, mustering politicians and representatives of 'civil society' to agree what form the transition to democracy should take, the political scene splintered. Nearly 400 parties formed. Some were dangerous, established by ambitious Big Vegetables who, to Mobutu's fury, now seized the opportunity to denounce the president and repackage themselves effortlessly as opposition challengers. But many consisted of no more than one loudmouth and his wife. To the farmers' parties, women's parties, lawyers' parties and handicapped parties, a new group was added: the so-called 'food parties' (partis alimentaires), ready to sell their CNS votes in return for a little sustenance.

Mobutu bought them as enthusiastically as he had once bought individuals. But in truth, he only needed to apply the odd nudge to ensure events at the CNS went in his favour. The squabbling of the 1960s swiftly resurfaced, fuelled by generous per diems which discouraged delegates from doing today what could be put off till the morrow. The main opposition parties fissured and split as, losing sight of their original ambition—a future without Mobutu—they quarrelled over who got to hold the lucrative post of prime minister. 'At one stage or another,' remarked Kitenge Yezu, the aide Mobutu entrusted with the job of undermining his opponents during that period, 'practically every member of the opposition ate at Mobutu's table.'

After one major CNS showdown, the country was for a while in the surreal position of boasting two premiers and two cabinets. One, reluctantly recognised by the West as being the real power in the land, was dominated by Mobutu sympathisers who cheerfully allotted themselves monthly salaries of $14,500. The other staged weekly 'cabinet' meetings in Tshisekedi's dusty compound and solemnly issued weekly round-ups of its pointless deliberations.

The country's political reform process stalled in its tracks. By applying his tried and tested techniques, Mobutu had triumphed once again. Tshisekedi retired to sulk in his compound, his credibility destroyed. Wounded to the core by the hostility displayed at the CNS—which had included proposals to drop the name Zaire and revive the flag and anthem of the Lumumba era—Mobutu nonetheless enjoyed the last laugh. He had demonstrated to a public that demanded democracy just how little they could expect from the men he had corrupted. The CNS, bitter Zaireans now joked, stood for 'Connerie Nationale Souveraine'—Sovereign National Bullshit.

Promising imminent elections in each New Year's address, Mobutu at the same time made it clear he had no plans to quit. 'I must complete my task,' Mobutu told the nation in 1994. 'I cannot leave this type of inheritance to posterity. Completing my task means leaving this country something worthwhile.' The leader who had once sent troops to burn down the printing presses now let the rag-sheets froth at will. They could say what they liked, he realised, as long as one basic principle was observed: he controlled the purse strings and commanded the military elite. Declared obsolete by Western diplomats in 1990, Mobutu was to succeed in stretching Zaire's so-called transition to democracy out for seven long years, transforming a temporary phase into a near-permanent condition. It was a remarkable achievement, by anyone's standards.

CHAPTER FIVE

Living above the shop

'We are partly to blame, but this is the curse of being born with a copper spoon in our mouths.'

> ***—Former President*** **Kenneth Kaunda,**
> ***commenting on Zambia's economic plight***

On my bathroom shelf, I keep a small rock taken from the edges of an open-cast mine on the outskirts of Lubumbashi. I went there a few months before it was captured by rebels and, not surprisingly, the mood was a little tense. I had spent the previous day at the headquarters of SNIP, the intelligence service, waiting for the local security supremo to spell out just how unwelcome a visit this was.

He had kept me pacing the corridors as long as was feasible, but had not gone so far as to expel me across the border, so by the next day I was peering into a kidney-shaped quarry, listening politely as a Polish engineer enthused over the ore being scooped from the site. The purity of the cobalt here, he explained, was 2.7 per cent, compared with 0.2 per cent in neighbouring Zambia. 'This place really is a geological scandal,' he sighed.

Whatever minerals my sample contains, they are clearly present in high concentrations. Around the reddish base rock, which crumbles in the hand with alarming ease, a knobbly blanket of dark green crystals is folded. The crystals have the feathery look of home-made ice-cream, and they twinkle in the light. Every time I pick it up I have to repress a smile. It brings to mind all the upbeat assessments written through the years by Canadian, Australian and South African mining companies trying to raise money for ventures in Congo. The ones that rave about the 50, 100, 200 million tonnes of copper/cobalt/zinc/tin/nickel just waiting to be extracted from this or that site, pooh-pooh doubts about the government's reliability and play down the small matter of a rebel uprising in the east.

The minerals are undoubtedly there, in concentrations high enough to make a metals analyst weep. But the rusting factories scarring Katanga's landscape, the abandoned yards, the stilled conveyor belts and dour expressions of the few technicians still at work are more accurate indicators to the province's prospects than any number of statistics-laden company reports.

They are all the evidence one needs that Congo has fallen victim to that paradox of sub-Saharan Africa, which dictates that countries with the greatest natural assets are doomed to war and stagnation, while nations with almost nothing somehow prove better at building contented societies. It is as though an impish god has decided to keep the scales of each country's destiny level: if one nation is blessed with oil, it will be cursed with a civil war, if another abounds in diamonds, they shall lie behind rebel lines, if a third is awash with copper, its leadership will prove too inept to organise its extraction. Or maybe the reason is simpler: the richer the nation, the more spoils there are to fight over. Sharing only seems to make sense when there is scarcely enough to go around.

There has never been a better example of the curse of natural riches than Congo. The mineral belt that fans out from Katanga's dry savannah into neighbouring Zambia contains copper and zinc in concentrations rival nations can only dream about and enough cobalt to corner the global market. Even the slag heaps looming over the decaying colonnaded towns built by the Belgians—Likasi, Kolwezi, Lubumbashi—could yield a fortune if reprocessed with modern techniques, so pure was the original concentrate.

Still waiting to be systematically charted, these were the deposits that made the Belgians so reluctant to lose control of Katanga, they encouraged Moise Tshombe to secede in the post-independence years. Nearly 500 miles to the north-west lies another gift of nature: the dark red gravel banks that trace the winding course of the Kanshi river, second-biggest source of industrial diamonds in the world.

There are other gifts: diamonds at Tshikapa in the south-west and

Kisangani in the north, what for a time was the world's main source of uranium at Shinkolobwe, and from the border with Uganda comes the enticing glitter of gold. Cadmium and cassiterite, manganese and wolframite, beryl, columbo-tantalite and germanium: metals with mysterious, evocative names. No wonder a US ambassador once memorably referred to 'the Congo caviar' in a cable back to headquarters.

Such natural wealth haunts the national psyche. Talk to any Congolese and at one stage he will cite his country's extraordinary attributes. 'We are a great country. No one has resources like us,' he boasts. Maybe it is an inheritance from the days when Belgian supervisors deliberately kept their Congolese subordinates away from strategic planning, or the hangover from a culture of slash-and-burn. But the investment, scientific know-how and marketing savvy needed to realise that potential are always glossed over. The mere existence of the assets entitles the nation to status and respect.

No one fell more thoroughly prey to the asset curse, the get-rich-quick fantasy, than Mobutu. For a president in constant need of ready cash, there could be little doubt where to turn. The 300 kilometre-long, 70 kilometre-wide mining concession Mobutu had forcibly wrested from Belgian control with the 1967 nationalisation of Union Minière du Haut Katanga (UMHK) and rebaptised Gécamines was the mainstay of the economy, accounting for up to 70 per cent of export receipts.

The Belgians had left behind a supporting network, an empire made up of mines, refineries, hydroelectric installations, factories producing anything from cement to explosives and sulphuric acid; town houses for its employees; schools and hospitals for their families; farms to produce food—even mills to grind flour—for the country's biggest single workforce: all the elements required to ensure Katanga was one of the world's most efficient copper-producing units.

So intrinsic did Gécamines seem to the nation's prosperity, Mobutu hatched the idea of the Inga-Shaba power line as a way of forever tying the mines to Kinshasa. Instead of relying on electricity from local dams, the plan went, this wonder of the world would render Gécamines—and Shaba—reliant on electricity generated 1,800 kilometres to the north, by the churning waters of the Zaire river. The power could be switched off at the touch of a switch by Kinshasa. The fact that the project, which involved fat commissions from the foreign companies bidding for the contract, resulted in a line which bypassed thousands of electricity-starved villages on its long route south was irrelevant. He wanted no more secession attempts.

In the healthy years of the early 1970s, with copper output hovering at between 400,000 and 470,000 tonnes a year and production of the far more valuable cobalt at between 10,000 and 18,000 tonnes, Gécamines alone could be counted on for annual revenues of between $700 million and $900 million. Until the world copper price collapsed in 1974, it must have seemed like a bottomless Horn of Plenty waiting to be emptied time and time again.

Mobutu's way of taking a cut was blunt in its simplicity. Sozacom, the state-owned subsidiary set up to market minerals abroad, would simply redirect a share of the foreign exchange Gécamines earned selling cobalt, zinc and copper on the international market to numbered presidential accounts held abroad, a practice coyly referred to by the World Bank and International Monetary Fund as 'uncompensated sales' or 'leakages'.

Another device used, according to officials of the day, was forward selling—mortgaging sales of copper and cobalt that had not yet been extracted. The proceeds went to the presidency, and the government would pay compensation to Gécamines to cover the gap in its accounts. Yet another trick was to exploit the margins between the various market rates for the metals, selling at one rate, logging another as the rate actually used for a transaction, and sending the difference to the presidency.

But often such subtleties were dispensed with. In 1978, an IMF official discovered that the central bank governor had ordered Gécamines to deposit all its export earnings directly into a presidential account. Two years later the practice had been only slightly modified, according to Steve Askins and Carole Collins, two US researchers who have investigated Mobutu's sources of wealth. Officials were stealing at least $240 million a year from Gécamines. In company reports the missing sums were logged under the wonderfully ambiguous term 'redressement exceptionnel déficitaire'—'exceptional deficit recovery'.

Once Zaire had acknowledged its economy was in trouble and promised to follow a route dictated by the World Bank and International Monetary Fund, of course, Gécamines came under close scrutiny. The Bretton Woods institutions were paying for the company's rehabilitation. Those in power were obliged to move sums from account to account to camouflage the missing sums. Cleophas Kamitatu, the man who sold the Japanese embassy, stumbled on one such operation in 1982, while serving as cabinet minister. The $100 million withdrawal from Gécamines' foreign exchange accounts threatened to scupper a Paris meeting at which he had hoped to win major promises of foreign aid for Zaire. 'The chairman of Gécamines told me the money had gone to Mobutu. I knew the $100 million had to be repaid into the Belgolaise bank or the conference would be a failure. So we filled the gap and made it look as though Mobutu had repaid the money when in fact we had simply borrowed the sum from Gécamines itself. So we got our foreign aid, but Mobutu got his money.' In fact a $100 million gap counted as fairly trivial. More typical was the $400 million which disappeared without explanation from Zaire's mineral exports in 1988.

In accordance with the Lord's Resistance Army principle, not all of this was going to Mobutu. Legal charges filed after Kabila's rebels took power give some insight into how well Gécamines' top executives were also doing during those years.

According to an indictment drawn up by the public prosecutor's office, the former head of Gécamines' commercial subsidiary unilaterally boosted his monthly travel allowance from an already hefty $15,000 to $30,000 during his final years in office, granting himself an additional $1,000 for every day spent off base. Setting aside several 'unjustified withdrawals' which ran into hundreds of thousands of dollars, this system alone allowed him to pocket 15.5 million Belgian francs in 1991 and 10 million in 1992.

Gécamines' huge network of associated activities also opened it up to abuse. The company acted as guarantor for state debts that went unmet, picked up hospital and hotel bills for its executives' relatives and sent its private planes shuttling across the country at their request. No wonder that by 1990 Zairean copper—so pure, so theoretically easy to process—actually cost nearly twice as much to produce as its foreign equivalent.

With the firm's receipts rarely making their way back to Katanga, there were no funds left over to maintain and renew the infrastructure left behind by the Belgians. Much of the equipment dated back to pre-independence and was constantly either out of service or being repaired. In the general climate of what is known in French as 'je m'en-foutisme' ('I don't give a damn'), managers began cutting corners. In the rush to get at the ore, underground tunnels were hurriedly excavated, their roofs held up with a minimum of props. In September 1990, the inevitable happened. The mine of Kamoto caved in, eliminating more than a third of Gécamines' output at a stroke.

The blows came in quick succession: a round of pillaging, echoing the anarchy breaking out up north; the departure of the company's experienced Kasaian workforce, expelled from Katanga in a bout of ethnic cleansing whipped up by the local governor and condoned by Mobutu, who wanted to send a warning signal to Tshisekedi, a Luba from Kasai, of how bad things could get for his tribespeople; and yet another orgy of looting.

But by then the company had already been crippled by a series of liberalisation measures that launched a new smuggling industry by making it legally possible for any Zairean to set himself up as a copper or cobalt dealer. 'Suddenly, everyone became a copper miner,' a white-haired Belgian manager, remnant of an expatriate workforce that once numbered 3,000, told me on a visit to Likasi's copper installations. 'The whole population began to steal from us.'

He had been in Katanga since 1960 and was clearly a member of that school too old to learn new codes of behaviour with the Africans who were once his country's subjects. Sitting in his dark office, he barked at his assistant to bring tea and expanded on the uselessness of post-independence government, which, he said, had not built a single house in the nearby town since the colonial power left. 'Everything here, the roads, the factories, the schools, was left by the Belgians.' As for the workforce that replaced departing white technicians, his racist contempt ran so deep it was no longer even tinged with anger. 'Give an African a job and he wants three wives, a nice suit and his status in society,' he said. 'But there's never anything to go with it. No commitment to the job in hand. Most of our workers have seven or eight children and they all have to be provided for. It's each for himself and devil take the hindmost.'

The corrosive scorn seemed a little more understandable when you considered what it must have been like sitting in that gloomy office year after year, witnessing the systematic cannibalisation of Gécamines by its own workforce. Having watched Mobutu and his cronies thoroughly milk the system, officials in Katanga saw little reason to hold back in the canter to self-enrichment. Lorries loaded with cobalt concentrate, officially labelled as 'tailings', were dispatched for sale across the nearby Zambian border with the benevolent collusion of local customs men. Vital equipment and spare parts were removed, peddled to operators in Zambia and South Africa who would then cheekily sell them back to Gécamines, the original owners. As one engineer acknowledged: 'We bought them twice.' But for this man the most outrageous incident came the day staff turning up for work discovered that 30 kilometres of high voltage cable supplying the plants had been snipped from the giant pylons during the night, presumably to be sold as scrap. 'The thieves had to switch off the power plant to do it, so the security forces must have been involved. It's not a job a small operator could have carried out.' By 1994 around a third of Gécamines' production was being smuggled south of the border.

In latter years, valiant attempts had been made to brake the thieving, said the Belgian manager. 'Customs men and generals were moved on and we set up roadblocks to stop lorries loaded with cobalt on their way to the border. But there is top level collusion in Kinshasa which means it continues, even if it's slightly less obvious.'

Less obvious, perhaps, because there was little left to steal. It took over thirty years, but by 1994, when copper production had sunk to 30,600 tonnes a year—less than a fifteenth of what it had been at its height—and cobalt output was 3,000 tonnes, the Horn of Plenty had effectively run dry. Revenue was zero. 'Gécamines,' in the words of Daniel Simpson, former US ambassador to Kinshasa, 'was as clean as a whistle. Mobutu had not only killed the goose that laid the golden eggs, he'd eaten the carcass and made fat from the feathers.' Gécamines was placed on what its chief executives described as 'a survival programme' and relieved of its crippling tax obligations. With the exception of the occasional quick-in, quick-out joint venture that barely scratched the surface of Gécamines' potential or its problems, the concession ground to a virtual standstill, with skeleton crews keeping the facilities ticking over in expectation of some far-off resurrection. In order to restore Gécamines' production to about 300,000 tonnes a year, the World Bank had estimated, any investor would have to assume debt in excess of $2 billion and invest another $1 billion.

Driving from one site to another, that figure seemed almost low. This was a landscape of quiet yards, empty skips, mysteriously dripping ceilings; plant after plant that looked as though its sole purpose was to breed rust in industrial quantities. I won't forget the vision of an overalled worker straddling a grating, pounding slowly with a hammer at a rock too large to go through a giant sieve. It was a job for an industrial crusher, but the crusher was out of order, so with a colleague holding the end of a rope wrapped around his waist to prevent a fall, he was reverting to the oldest mining technique known to man. Watching him sweating, I was reminded of a joke told south across the border. 'What did Zambia use before candles came along?' it goes. 'Electricity.'

But it was at the Shituri plant that the extent to which Gécamines had lost the fight against its own staff became clear. Against all odds, cobalt was still being processed here, electrolysed on large plates suspended in solution and then noisily ground out—dark granular fragments dropping into a large plastic container that looked suspiciously similar to the kind of sacks normally used for wheat flour. A South African security company had been hired to patrol the grounds and 250 armed guards monitored a plant working, in any case, at a mere 25 per cent of capacity. Yet management had still felt the need to store the sacks of cobalt in a padlocked warehouse, stretch fronds of barbed wire across the roof and plug the whole apparatus into the national grid. 'That way, if anyone tries to be clever by going in by the roof, they get fried,' chuckled a Gécamines worker, drawing my attention to the skull and bones sign on the warehouse door. 'If there was a world competition for breaking and entering, we Zaireans would always win first prize.'

Many Katangans believe they have been made to pay the price for their autonomous leanings, which simmer on today. 'They had to make Shaba poor, so that we would be dependent on Kinshasa. The destruction of Gécamines was deliberate,' insisted a member of a local political party. It is a conviction that leaves an abiding sense of resentment. 'If the Zairean economy lasted as long as it did, it was thanks to us. For thirty years they bled us dry and in exchange, what did we get? We were colonised a second time, first by the Belgians, then by the Kinshasa regime.'

But Mobutu's next stop in the hunt for disposable income was to be made on different terms. When Gécamines, backbone of Zaire's economy, showed its first sign of faltering, he turned his attention north-westwards, to the rebel province of east Kasai, a region with just as many reasons as Katanga to long for autonomy.

East Kasai is home to the Luba people. Dubbed the Jews of Congo, the Luba are regarded with suspicion by their fellow nationals as a little too good in business; aggressive wheeler-dealers overly prone to share the fruits of their worldly success exclusively with their tribesmen while ruthlessly boxing out other ethnic groups. 'Let one Luba into your business, and next thing you know, they'll be running it,' Kinshasa residents warn.

Etienne Tshisekedi, Mobutu's most formidable challenger until events in Rwanda rendered him irrelevant, hailed from there and having him as their champion always underlined the Lubas' status as outsiders. Only curmudgeonly Kasai, one feels, could have got away with the step community leaders took in 1993, when they took against a new currency issued by Kinshasa. Deciding, with considerable justification, that this was a monetary scam designed to line the pockets of politicians in the capital and was bound to have an inflationary impact, the local elite simply decided to boycott the new notes and stick with the 'ancien zaires' of old.

For the diplomats in Kinshasa, it was a move which, if left unchecked, would raise the interesting question of whether Zaire was still a state, or simply an empty space defined by nine countries' frontiers. 'It was a bit like Yorkshire unilaterally deciding that from now on, it was going to use Monopoly money,' said one. 'It raised some fundamental questions about what makes a nation a nation, and what it was exactly that Mobutu thought he was presiding over.' Yet Kasai was allowed to do its own thing for five long years.

By this late stage of his regime Mobutu was no longer interested in symbols of sovereignty. He needed cash, and the readiest source, once copper had lost its sheen, was to be found in Mbuji Mayi, the town built where a subsidiary of a subsidiary of the great Congo river meanders lazily across the plain, depositing the tiny stones washed from the green-grey pipes of kimberlite, embedded—in the analogy favoured by mineralogists—like so many giant carrots in the soil.

As early as 1907, a colonial prospector had taken what looked like an interesting pebble from this site for analysis in Europe. It lay ignored until an engineer preparing another trip started sorting through samples, stumbled upon it and confirmed that yes, this really was a diamond. By now no one could remember where the sample had been collected and it was only in 1913 that Kasai's status as a diamond region was confirmed. Since then, diamond fever, as dramatic in its social effects as any Klondike Gold Rush, has held the area in its grip. So valued are these deposits, you need special permits to visit the region and at regular intervals the government in Kinshasa, convinced too many outsiders are getting their share of the booty, ejects the Lebanese middlemen who home in on the province like sugar-crazed wasps.

Mbuji Mayi is a curiously soulless settlement, with no tangible centre, boasting none of the elegant civic buildings left behind elsewhere in Congo by the Belgians. It is a purely functional conurbation, dedicated to making money, with little left over for less focused activities. Its pulsating artery is the Avenue Inga, the rutted, muddy road on whose white-washed walls have been painted, in garish colours, the names of the buyers within, a promise that only they will offer a fair price and—as an example of what is expected from those crossing the lintel—a sketch of a sparkling, blue-white diamond worthy of a monarch's crown.

Inside the carefully guarded compounds, tough young Brits working for the local branch of the South African giant de Beers, homesick Lebanese middlemen who miss their families, and sometimes—but not often—Congolese themselves, sort delicately through piles of what look like coloured sugar crystals. They are searching, in this region where industrial-quality diamonds are the norm, for that rarity: a fault-free, gem-quality crystal structured to slice cleanly under the cutter's blade and guaranteed to fetch thousands of dollars in Antwerp.

'Fortunes can be made here,' said Ahmed, a Lebanese buyer. 'But you really have to know diamonds. And you have to be willing to wait months on end for something special to come along. A lot of people don't have the patience.' He displayed his latest cache lovingly on the back of an envelope. 'These are mostly white stones, which is unusual for here. I spent $30,000 and they should fetch $40,000 in Antwerp. You have to make at least a 10 per cent margin, as you have a lot of expenses to cover.'

In a perfect world, very few of the sixty-odd diamond comptoirs (counters) in Mbuji Mayi would be operating at all. For the diggers they buy from, ragged men who spend their days waist-high in the

Kanshi river, sifting red gravel through crude sieves, taking cover when they hear the sound of approaching voices, are what the Angolans call 'garimpeiros', or illegal diggers. These freelancers work land nominally awarded to the Société Minière de Bakwanga (MIBA), the company in which the state owns 80 per cent and a Belgian company has the remaining 20 per cent. It is a 5,000-square-kilometre mining concession the company has neither the facilities nor the energy to police. 'Ninety-five per cent of the diamonds being bought by the comptoirs come from our concession,' estimated a MIBA official. 'I don't think there's a single legal counter in Mbuji Mayi. But if you crack down on the comptoirs the diggers will just take the stones into Angola and sell them there instead.' Respecting the regulations, after all, has never been the norm in an industry in which the value of diamonds smuggled out each year far outstrips the $300–400 million going through official, and therefore taxable, channels.

Diamonds were always going to be the perfect product for Mobutu: tiny, easily smuggled across borders in a briefcase, handbag or pocket and needing none of the clumsy apparatus of electrolysers, smelters and refiners or the messy infrastructure of rickety railways, road haulage and quaysides associated with base metals. He could play the same game as the buyers on Avenue Inga, setting up diamond counters which massively underestimated the value of their wares in official declarations and then unloaded the garimpeiros' produce onto the Antwerp market. But Mobutu could also go directly to the biggest diamond digger of them all and demand a presidential share. And in Jonas Mukamba, the long-standing government representative running MIBA, Mobutu had a man he could do business with.

A tall, imposing Luba known for his tendency to speak his mind, Mukamba had been a player on the national scene as long as Mobutu himself. Engraved on the memories of most Congolese is the fact that it was Mukamba who accompanied Lumumba and his two fellow prisoners on their terrible flight down to Elizabethville in 1961, handing the former prime minister over to the men who would take him to his death. As with Mobutu, the oily stain of fratricide has clung to MIBA's president for more than thirty years, lending him a sinister glamour which has done nothing to blight his progress through the world.

Maybe this shared complicity made cooperation between the two men easier. For when Gécamines' top executives stopped being able to deliver, Mukamba stepped in. 'Mukamba was probably creaming off between $1.5 and $2 million from MIBA for Mobutu each month,' guessed a former government economist. 'People often say the advantage of a private company is that it puts a stop to this kind of thing. Well, MIBA's mixed economy status made not the slightest bit of difference.' Mukamba obliged Mobutu in other ways. Foreign dignitaries the president wanted to impress were taken to the MIBA building. The company's security guards would be ordered out, each visitor handed a shovel and sack and invited to help themselves to MIBA's raw diamonds. If today the company officially refuses to comment on its past, Mukamba's role as Mobutu's fund-raiser is quietly acknowledged by staff who took over administration after the president's departure. 'But quite honestly, what else could he do?' asked a MIBA colleague. 'It's easy to criticise, but it's not black and white. Towards the end it was impossible to work with Mobutu if you didn't play the game. And in the process Mukamba did a lot of good to the town.'

For, in exchange for the presidential tithe, Mukamba was left to run Mbuji Mayi and its environs largely as he pleased. Filling the space left by the absent government, he turned it into the ultimate company state. MIBA repaired the roads, pumped drinking water, supplied the town with electricity and sold food at subsidised prices. MIBA contributed funds to the nascent university and paid for foreign professors to fly over to teach. A young Kasaian could easily spend his life in MIBA-funded institutions: living in a company house, attending a MIBA school and dying in a hospital supplied with MIBA drugs. 'I'm a businessman, but I'm also a politician and my job is to look after the population,' Mukamba told me shortly before his removal. 'The people are very aware of what MIBA is or isn't doing and if we don't do it we are severely criticised.'

With its own scrip and a company-funded health and education system, bolshy east Kasai enjoyed self-rule of sorts. But it was always autonomy by default, and it never touched great heights. By the late 1990s, despite MIBA's repair work and its electricity plant, many of the main roads still subsided in the rain and much of the town was plunged into darkness at night. The much-vaunted university was little more than a couple of outhouses and a dusty roomful of books. A MIBA-sponsored orphanage was a bleak, furnitureless shack, where runny-nosed children could recite the catechism but went without shoes. Residents had accepted their half-baked secession as the best deal they were likely to win and got on with the task of making money, an approach that infuriated the more idealistic amongst them. Gaston Muyombo, a Catholic priest, blamed the approach on what he called 'Bantu philosophy': 'People cling to life and are not yet at the stage where they will fight for the quality of that life. They feel as long as they are surviving, that is enough.'

Mukamba's freedom of movement was curtailed not only by his role as unofficial presidential cash provider, but by the fact that MIBA was still paying cripplingly heavy national taxes in Kinshasa. 'Our dearest wish would be to spend those taxes locally. But the law obliges us to pay,' he explained. 'For more than thirty years, power has been far too centralised in Zaire.'

Combined, the two depredations helped send MIBA down the same sad route as Gécamines. By 1997, the year of Mobutu's downfall, output had fallen from 10 million carats a year to 6.4 million. The company had been loss-making for six years, and it too had been put on a specially lightened tax regime. The curse of prosperity had struck again. Yet another thriving national industry, on which another state might have built a vertiginous rise to international prominence, had been sabotaged.

The shift of focus from Katanga to Kasai, from copper to diamonds, marked another stage in Mobutu's itinerary. The man who had founded his empire on his ability to distribute sweeteners was now being outstripped by the more enterprising members of the political class he had helped create, who had learned their lesson a little too well. Several of Mobutu's sons, his personal aides, the generals, were all soon running their own diamond-buying counters on the Avenue Inga. They doubled as conduits for the higher quality gem diamonds being mined by the UNITA rebel movement across the frontier in Angola, which needed legitimate commercial outlets for the stones to fund its military campaign. 'Mobutu was losing his capacity to rein those guys in,' said a US Treasury official. 'The illicit diamond counters were wandering out of his reach and his ability to plunder the various state mechanisms had shrunk enormously.'

In his role as diamond expert, Larry Devlin tracked the same phenomenon. 'I heard through my contacts that one of Mobutu's closest aides was going to Antwerp with $9 million worth of diamonds to sell on the president's behalf. He'd tell the buyer, "Give me a receipt for six and I'll take the other three." In the old days that would never have happened. He would have gone back to Mobutu with nine and waited for the president to give him his share. He would never have dared take his own cut first.' The kleptocracy was no longer the creation of one man. It had acquired its own unstoppable momentum.

On one of the main avenues of Kinshasa's tree-lined Gombe district, home to ambassadorial residences and ministries of the city, a mastodon of a building constructed in the shape of a giant reversed 'C' lies behind high walls of concrete and iron. It used to be possible to drive straight past this cement hulk, but ever since a bout of shooting between two rival army units alerted the new authorities to the institution's vulnerability, traffic has been diverted down narrow side streets. But from this distance you can still spot a curious white sphere high up on one of the building's corners. This used to be where a mosaic of Mobutu, complete with leopardskin hat and dark glasses, surveyed the scene. It was whitewashed over in the days when Kabila's rebels walked into town and, in true Vicar of Bray style, the capital's residents rushed to 'rebuild their virginities', in that magnificent French phrase.

But it is easier to paint over a portrait than to cancel out the past. It was here, at the central bank, that the final scenes in Mobutu's kleptocratic system were played out. Having reduced Gécamines and MIBA to shadows of their former selves, with no substantial revenues coming in, the president and his increasingly wayward elite were left with the option of last resort: printing money to survive. The presses would be ordered into action, army lorries sent to the central bank and the thick wads of pristine zaire notes quietly unloaded on what Radio Trottoir had dubbed 'Wall Street': the alleyways where scores of Kinshasa's divorcees, widows and single mothers—these feisty moneychangers were nearly always women—would sit with bulky bags of money on their knees, setting the day's exchange rate. Dumping their notes, still in the plastic wrappers in which they had been issued, the top officials would hurriedly exchange them for dollars, Belgian francs or French francs, the only stable landmarks in a world of constantly shifting value. But as word spread on Wall Street of yet another mystery delivery, the day's rate would change and the zaire would fall—and fall.

Inflation, which had reached double-digit figures sufficient in themselves to bring down an accountable Western government, suddenly rocketed in 1991 to a mind-boggling 4,130 per cent. The next year it fell slightly to 2,990 per cent. But the next year it was back up to 4,650 per cent and in 1994 came the worst of the worst: inflation ballooned to 9,800 per cent.

For Zaireans paid in local currency, the effect of what was effectively an unofficial tax on every financial transaction was disastrous. In the time it took to drink a coffee, the rate could have changed a couple of times. Dither too long over the bill, and it might have to be altered. Return from a long trip and the store of zaires that had bought a family a meal before you left now scarcely afforded a bar of soap.

In supermarkets, no one bothered ticketing goods individually any more, so quickly did the prices change. Instead they were classed in categories with a single index, easily updated, giving that day's price for each category of goods. Each individual note was now worth so little, the banking industry effectively ground to a halt, unable to muster the liquidity needed for major transactions. Sometimes, behind the tellers, you would see hillocks constructed of soiled, strangely aromatic zaire notes, stacked against the wall in brick-like blocks: destined for some small business struggling to pull together the salary for its workforce, perhaps, or to buy a photocopier. There was rarely enough cash for anything more ambitious. Checking the amount could take hours, despite the fact that to simplify counting, notes were split into convenient 'paquets' of twenty-five. You trusted your black market moneychanger, in fact you trusted every Zairean you dealt with, not to subvert the entire system by sneaking a couple of notes out of each paquet. Ironically, a crisis created by such top-level dishonesty bred its own moral norms amongst its victims, adhered to with a greater degree of conscientiousness than the rules of a conventional financial system.

Indeed, it gave birth to an imaginative mutual aid system amongst those still tenacious enough to want to operate in a society where the banks had become irrelevant. Coffee exporters and arms traders, aid organisations and diamond smugglers found themselves strange bedfellows as they established an informal money-trading network. A single phone call would enable a factory boss to locate the zaires needed to pay his work-force, or a Lebanese dealer to find the dollars he needed to buy his diamonds. It was do-it-yourself banking and it worked. 'I find it quite inspirational,' a British businessman once confessed. 'Hundreds of thousands of dollars worth of currency will be traded over the phone, the transaction takes seconds to go through, rather than the weeks involved if it were being conducted through banks, everything is done verbally and no one ever welshes on a deal, because they know if they did the whole apparatus would collapse around their ears and everyone would lose out.'

But even if the process was taking place in graceful slow motion, the system was indeed imploding under the weight of its own eccentricities. Each time a new denomination was issued in a forlorn attempt to keep up with inflation, politicians would wait with bated breath to see if the population would accept it as legal tender or refuse it as inflationary. That was the step that helped push the soldiers to riot in 1993, when they found their wages being refused in shops. One of the last bills issued under Mobutu—the 500,000-zaire note cheekily dubbed the 'prostate' in honour of his afflicted organ—was rejected *en masse* in Kinshasa, providing a few mouvanciers with a wonderful opportunity to exploit. Appropriating notes rendered worthless in Kinshasa, they chartered planes and flew stacks of prostates down south to Lubumbashi, where they dumped them wholesale onto a more amenable black market.

In Kasai, of course, the new zaires had never been accepted at all, presenting a lucrative opening for officials who hoarded 'ancien zaires' instead of burning them as directed, then offloaded them in the Kasaian trading centres of Mbuji Mayi and Kananga. In the far east, new zaires were accepted but traded at a different rate against the dollar from Kinshasa, another opportunity for those lucky enough to travel to make a profit on the spread. One country, at least four separate currency zones: Zaire was beginning to crack at the seams. Mobutu the African nationalist may not have liked all that this implied. But given his own past, what could he do about his entourage's increasingly reckless freelance activities? Moral sermons would have come across as unacceptable hypocrisy. 'Since he was taking himself, he could not punish others,' said Kitenge Yezu.

None of this world of Alice-in-Wonderland finances, of course, made its way into the monthly and yearly reports issued by the central bank, where accountants and economists painstakingly massaged the figures, constructing a sophisticated simulacrum of financial respectability that might, at a pinch, fool some IMF or World Bank expert still interested enough in Zaire to ask to see the books.

As Mobutu's options narrowed, Western governments pinned their hopes of reform on Kengo Wa Dondo, the light-skinned, razor-sharp former attorney-general who had served twice as prime minister during the single-party era, before once again being nominated premier in 1994. The dominant figure in a group of Big Vegetables to emerge in 1990 as pitiless critics of one-man rule, Kengo was, the embassies believed, acute enough to realise the financial mismanagement had to stop. Kengo did enjoy initial success in bringing down inflation. But then came a series of scams so outrageous, so ambitious, they betrayed an utter disdain for the law, the population and the very tenets of a nation state by those involved.

In the early hours of 2 September 1994, a Boeing 707 belonging to a Liberia-registered company landed at Ndjili airport. The plane, which was inspected before taking off for the interior without authorisation, proved to be carrying an extraordinary load: 30 tonnes of banknotes, amounting to 12–15 billion new zaires. Had a forgery ring been exposed? The truth turned out to be rather more complex.

So strapped for cash was the Kengo government, it had resorted to entrusting the printing of its new currency bills, carried out in Argentina and Brazil, to Lebanese intermediaries. They would fund the issue and then be paid for services rendered out of the new banknotes that resulted. But the well-connected Lebanese businessmen involved, Naim and Harif Khanafer, went a step further. Exploiting their unique position as agents of the Zairean monetary authorities, they asked the Brazilian and Argentine companies to issue not just one copy of each numbered bill, but two notes, three notes, maybe four. Identical in quality and detail to the original, these were not technically speaking forgeries at all, and were impossible to isolate. Dumped in their tonnes on the black market, their effect was bound to be catastrophic.

After an emergency cabinet meeting, Kengo's information minister went on national television to explain and denounce the scandal, relieved, one suspects, to be able to attribute inflation to something other than money-printing authorised by the Treasury and the central bank. The air company's licence was revoked, the Lebanese brothers detained for questioning, Interpol's help was requested and a top-level legal inquiry was launched, with government investigators dispatched to Brazil, Argentina, Belgium and France. One month later, a second cargo of 14 tonnes of banknotes was discovered in the riverside town of Mbandaka, and this time the government managed to confiscate the load.

No one will ever know who was behind the action taken by the Khanafer brothers, but what is certain is that as Lebanese nationals reliant on their contacts to work and live in Zaire, they would never have launched an operation of this criminal grandeur without high-level patronage. Asked to name those responsible, Kengo cited certain 'civilian and military individuals' but declined to go into any detail 'so as not to prejudice the legal investigation', he said.

With weary predictability, the inquiry was quietly allowed to subside. Legal charges were never filed against any of those responsible and in December of that year the government, perennially short of cash, actually ordered the freeze on the 14 tonnes of Mbandaka banknotes to be lifted so it could pump the 'forged' bills into the money system. Kengo had underestimated the strength of the 'subterranean forces' behind the scam. Confronted by the powerful lobbies involved, he had preferred political survival to a showdown that might lose him his premiership.

The ugly face of a regime that was sucking the lifeblood from its own citizens had been publicly exposed. If they were slightly taken aback at the sheer depth of the greed exposed, Zaireans were certainly not shocked by the motives themselves. They had grown accustomed to the notion of state as ravaging predator. What mattered was knowing how to cope with the outcome.

CHAPTER SIX

A nation on Low Batt

Mobutu, French President Jacques Chirac and Bill Clinton are all on the same plane, returning from an international conference. Halfway into the flight, the pilot announces that he has lost his way in the fog and has no idea where they are. Clinton opens a porthole a few inches, reaches down and feels around. 'I know where we are,' he announces. 'We're over the US.' 'How do you know?' ask the other two. 'I just felt the top of the Statue of Liberty.' A few hours later, and the pilot is still lost. Chirac opens the porthole and reaches down. 'I know where we are. We are over France,' he says. 'How do you know?' the others ask. 'I just touched the Eiffel Tower.' Several hours later, the plane is still lost. Finally Mobutu rolls up his sleeve, opens the window and reaches down. 'I know where we are,' he announces, withdrawing his hand. 'Where?' 'We're over Zaire.' 'How can you be so sure?' ask the other two. 'Someone just nicked my Rolex.'

Joke popular in Kinshasa's expatriate community

By the mid-1980s, Zaire's Belgian-installed telephone network had disintegrated to a point where communications—both internal and international—were becoming impossible. It was then that a young American who had recently lost his job at an airline office came up with the bright idea of issuing Kinshasa's movers and shakers with Motorola radio sets which allowed them to keep in touch with each other within the city limits.

Not long afterwards a private cellular telephone system was set up and the Motorolas were replaced by chunky mobile phones. And so Telecel was born, an example of how a collapsing state structure could be sidestepped or simply substituted when the needs of the elite became acute. Road non-existent? Buy a four-wheel drive. National television on the blink? Install a satellite dish in your back garden and tune in to CNN. Phone out of order? Hire a Telecel. As Zaire crumbled, one community, at least, could afford to buy its way out of anarchy.

Customers might moan about the crippling seven dollars a minute the company at one point charged for international calls, but they were careful not to be cut off. Long before mobile phones became the rage in the West, owning a Telecel in Zaire was the ultimate prestige symbol, the difference between being a player and remaining on the periphery. For the new arrival, whether diplomat or journalist, it was a convenient way of sorting out the sheep from the goats. If, at the end of your meeting, you discovered that your interlocutor did not own a Telecel, you knew that no matter how worthy or articulate, he bore the unmistakable stamp of irrelevance. In contrast, I knew I was in the presence of greatness when I watched Bemba Saolona, Zaire's leading businessman, juggling a row of Telecels lined neatly up on the coffee table in front of him as they trilled in swift succession.

As with all things in Zaire, the Telecel and its idiosyncrasies became part of the language. One of the problems with a Telecel was the speed with which its rechargeable batteries would expire. As they ran down, the words 'Low Batt' would flash up on the display, accompanied by a two-tone bleep that would become more and more insistent until the line went completely dead. 'Je te rappelle, je suis Low Batt' ('I'll ring you back. I'm Low Batt') users would warn their callers. As time went by, the phrase took on something of a symbolic meaning. By the 1990s, the entire nation seemed stuck in a permanent state of 'Low Batt', surviving rather than living, ticking over without ever flaring into life.

The Zaireans had developed their own language to deal with this depressing reality, ironic word games replete with scepticism, the only form of quiet rebellion on offer in a system seemingly impervious to change. Cock-ups were attributed to 'Facteur Z', the Zairean factor, delays to 'Heure Zairoise', the lethargic local timescale. The capital once known as 'Kin-la-belle' was now dubbed 'Kin-la-poubelle' (Kinshasa, the rubbish dump), testimony to the mountains of garbage collected but never taken away. The men who sold stolen petrol on the roadside were known as 'Khadafis', in tribute to Libya's oil-rich president, while the urchins who slept on the streets were called 'phaseurs' (Lingala slang for 'sleepers') because, a friend joked, 'they were in phase with life'. The unemployed young men with nothing to do but stand on street corners discussing topical issues were scornfully dismissed as 'parlementaires debouts' (standing parliamentarians). Ask one of these how he was doing and the answer would never be the automatic 'bien'. 'Au rythme du pays', (in time with the country) he would reply, with a shrug, or 'au taux du jour' (at the day's rate), a reference to the national currency's unstoppable decline.

Returning from trips abroad, I never ceased to be amazed by how much further batteries I had assumed to be already near-exhaustion had sunk. I could measure it in the state of the taxi I hired on a daily basis and the mood of its owner François, the grumpiest, if most resourceful, driver in town. When I first arrived, his eighteen-year-old Fiat, a cast-off from a grateful Neapolitan businessman, was already nearing the end of its natural life. Too often, it needed to be push-started to coax it into action. The triumph of determination over logic, its seats had been eviscerated, its windscreen cracked in two places. Long gone were side mirrors, indicator lights, horn and windscreen wipers, which made driving in the rain particularly exciting. The electrically powered windows had to be manually heaved from their slots, the engine only started when three wires were twined together. At puddle level, the rusting bodywork was developing the delicate texture of lace. Front doors had a disconcerting tendency to fly open at high speed and sometimes had to be tied to the chassis with a rope made from knotted plastic bags.

Never light, François's spirits deteriorated in tandem with his car. Aware that an extra scratch or dent would now make little difference to the Fiat's roadworthiness, he paid only lip service to traffic regulations. His response to criticisms of his careless habits had become so aggressive, shocked bystanders would come to complain when I got in. As we clattered along at thirty miles per hour, the exhaust pipe occasionally trailing along the ground, François, who was plagued by stomach ulcers, would sit muttering darkly to himself: 'This country is screwed,' shaking his head. 'Who would want to buy it? Not even the Japanese. If I'd had any sense I'd have left long ago.'

The extent to which the nation was running on empty really came home to me during my regular trips to the Ministry of Information. The ministry was located on the nineteenth floor of a concave tower built, during the sweet days of international cooperation, by a company run by the then French President Valéry Giscard D'Estaing's cousin. Across the continent, I have trudged, cursing, up the urine-scented stairs of such high-rise monoliths, dreamed up by men who modelled Africa's itinerary on their nation's own. Seemingly incapable of conceiving of a future worse than their hopeful present, the foreign engineers could not imagine the day when electricity would be spasmodic, spare parts impossible to find, maintenance a joke. But so it proved and there the hulking anomalies sit: twenty-storey cement monstrosities caught out by history; designed for air-conditioning, wall-to-wall carpeting and smoothly operating lifts; marooned in countries heading back to pre-colonial times.

At the Ministry of Information it was always worth sending a scout ahead of time to find out whether the lifts were working that day. In any other country, you might also call ahead to find out if the man handling your documents was at his desk. But as a low-ranking Zairean official he did not boast a Telecel. One option was to stand at the bottom of the building and shout upwards until someone above heard your calls and established that your man was there. If you were unlucky, you then faced a painful climb up the dark stairwell, which doubled as a male lavatory. If you were lucky, tapping on the first-floor metal doors with a pair of keys would alert someone at the top to your presence, and they would send the lift down to collect you. Occasionally, the lift would be working but the lights were not. The solution on such days was a workman's lamp, solemnly handed from one passenger to another as they entered and left the lift.

At the top of the Ministry, there was always a refreshing breeziness. The air-conditioning did not work, but the windows were kept open and the air wafted through the building. Here you got a bird's eye view of the world. Toy cars could be seen driving along the Avenue 24 Novembre, people had become the size of tiny dolls and the layout of the nearby military barracks, the People's Palace and the modernistic sports stadium—two more foreign architectural gifts to a grateful President Mobutu—took on a new dimension.

But the detail that always stuck in my mind was a mundane one.

From up there what went virtually unnoticed at ground floor level emerged: a pattern of neat squares carved into the red earth. For kilometres around, all open spaces had been divided into carefully watered plots. On road verges, traffic islands, what should have been the lawns of the ministry itself, the distinctive spiky leaves of the cassava plant grew. This was a green city, but it was not greenery aimed at pleasing the eye. While Mobutu amused himself landscape gardening in Gbadolite, a nation on permanent Low Batt had no time for lawns. Preoccupied with the immediate problem of getting enough to eat, the residents had turned Kinshasa into one massive vegetable allotment.

To the north-east, the outline of a walled institution on the edge of the city centre could just be glimpsed. Constructed by the Belgians, it was a low-lying building whose main entrance was guarded by two men in beige uniforms. At street level their task seemed a tedious one—opening and closing the heavy metal gates with mind-numbing frequency. But closer investigation revealed the two to be armed with the kind of black rubber truncheons used by riot police. Their gaze was watchful and they scanned the crowd filtering through the entrance with care, on the lookout for inmates making a break for freedom.

These were no watchmen on duty at Kinshasa's Makala prison. This was the hospital once known as Mama Yemo and now called Kinshasa General Hospital. Their task was to physically restrain patients foolhardy enough to try and abscond without meeting their bills. For the cash-starved administration of the city's main hospital, non-payment was a luxury it simply could not afford. 'We call it impounding the ill ('séquestration des malades'),' explained Dr Henri Kasongo, head of emergency surgery at Mama Yemo. 'All the hospitals do it, although they'll never admit to it. The guards have to be very alert, sorting out those who have been sick from those who are well. Often people will try to blend amongst the public at visiting time. There are lots of escapes because, to be honest, these hospitals have become like prisons.'

Equipped with 2,000 beds, Mama Yemo for a long time claimed the proud title of central Africa's biggest hospital. Mobutu clearly had a glorious future in mind when he named it after his mother and ordered a bronze bust of her to be placed on one of the paths running between the blue and white painted pavilions. As the 'people's hospital', it was supposed to receive almost 50 per cent of the health budget, but the money never followed the good intentions. Supplies dwindled. Salaries, on the rare occasions when they were paid, fell to laughable levels. Sick mouvanciers would check into private clinics or, following Mobutu's example, fly to Switzerland in their private jets for treatment. Mama Yemo was left to fend for itself.

It was a situation, said Dr Kasongo, a tall man whose prominent jaw hinted at a certain pugnaciousness, that had led ineluctably to a condition not covered by the traditional medical textbooks: an overall hardening of the heart. 'In the popular press we are portrayed as butchers, coldhearted and utterly ruthless. But without the crumbs we get from the patients the hospital would close down completely. So what is the alternative?'

From the outside, the situation did not look too bad. White egrets picked their way across the green lawns, looking for edible rubbish. Washed clothes spread on hedges to dry provided bright patches of colour. There were the usual gaggles of women bringing food in metal containers to their loved ones and signs banning rifles, prompted by one too many incidents involving wounded soldiers demanding priority treatment. A bullet hole in one wall bore witness to how insistent they could become.

But inside the wards, it became clear why even the staff openly referred to Mama Yemo as a 'mouroir' (death chamber). Doors were splintering, the walls badly needed painting and there was no window netting to ward against malarial mosquitoes. Men and women, soldiers and civilians, suspected AIDS-carriers and the HIV-negative were mixed indiscriminately together, their narrow beds only inches apart. The air was pregnant with that acidic aroma you rarely notice in Western hospitals, where it is swamped by a layer of soap and disinfectant. A mixture of pus, warm flesh, urine, human secretions.

With its promise of possible infection, the sweetish smell clung to the hairs of the nostril, hovering hours after its source had been left far behind.

When it came to emergencies, heaven help those unlucky enough to be admitted *in extremis*, without friends and family on hand to offer instant cash. At one stage, the hospital used to keep a stock of drugs and supplies. But they vanished as patients supposed to reimburse the hospital once the emergency was over proved unequal to the task. 'Now the doctor is in a dilemma—the patient has no money, the doctor sees that he is dying but he has no drugs or supplies. I probably lose two out of every ten patients admitted with serious problems that way,' admitted Kasongo.

Those with more time at their disposal knew what to expect: scalpels and sewing thread, plaster of Paris and rubber gloves—all must be brought before the doctor would lift a finger. And at the end, the guards at the gate were instructed not to let the patient out until what was delicately termed the 'service de recouvrement' (recovery service) had ensured costs were covered. 'A patient may end up owing a couple of thousand dollars. But how can a civil servant who is not being paid his salary afford that? So he will be physically prevented from leaving and usually after a month or so his family will have gathered enough money together to get him out.'

Those most vulnerable to such pressure were young mothers, whose new-born babies could quietly but firmly be kept in their cots till the bill was settled. 'One woman spent two months here,' recalled Kasongo. 'She had given birth and couldn't leave. Everyone was laughing about it.' A rival hospital, he said, had gone so far as to set up a special room for patients with outstanding debts as a way of streamlining the problem. 'It's more convenient, as it frees up space. As a doctor, I don't want to have wards full of healthy people.'

Nothing could be taken on trust. At times the hospital acted as a pawnbroker, confiscating radios, watches or televisions as surety. 'Patients will often claim they have nothing, that they are destitute. In that case, they will be asked if they have any possessions. If one admits, say, that he has a television, we'll say, "OK then, bring that in." '

Even death, it emerged, offered no escape from debt. Just as living patients could be held hostage, bodies could be retained in the morgue until the family settled—a practice, the doctor acknowledged, that violated every moral precept in a culture where ensuring that a body was properly laid to rest had always been of enormous spiritual importance. 'Impounding a body is scandalous, a monstrosity, it's simply unacceptable. Those who can will pay up immediately to get the body out of the morgue.' But not everyone could. Hence the growing tendency for poverty-stricken families to abandon their relatives' bodies. Every three or four months, Mama Yemo's morgue filled with unclaimed corpses. Given the hot climate and intermittent electricity supplies, its refrigerators usually proved unequal to the task and the bodies began to decompose. At a certain point, when the problem became overwhelming, volunteers with strong stomachs would remove the corpses and bury them in a mass grave. National television sometimes broadcast images of these heroes at their terrible work. Behind their face masks, they retched as the corpses fell apart in the process of being lifted, shedding hands, feet, limbs.

Such horrors helped ensure that patients still waited until all other options were ruled out before admitting themselves to Mama Yemo. Doctors, for their part, avoided surgery unless absolutely necessary, for fear of infection, aware that poor medicine was encouraging diseases once beaten into submission to flare back into prominence. 'Sleeping sickness, leprosy, typhoid. All these illnesses that were nearly under control are coming back, and it's not a pure coincidence.'

Kasongo himself was entitled to a $20 monthly salary, but he had not been paid for five months. Most doctors in the state sector, he said, dreamed of moving abroad. He was one of the few who wanted to improve conditions here, despite his scepticism. His hair was thinning, he was no longer a young man. Sometimes, he acknowledged dourly, he wondered whether he had made a mistake choosing to become a doctor at all. 'I didn't study medicine to cure people on the basis of wealth, to extort money from patients or to let people die without intervening. I get demoralised. If I met someone who wanted to go into medicine today I'd say, think twice. It's a big disappointment. There's something sub-human about this way of life.'

As we walked towards the exit and the uniformed sentinels, standing like Cerberus at the gates to Hades, we paused at the plinth which once supported Mama Yemo's bust. Within hours of Kabila seizing power, the bronze had been toppled by members of staff. Maybe they recalled another symbolic statue-felling, staged at the bidding of the leader they then believed in, who left his mother's namesake to rot. 'It's funny how an event will suddenly reveal what everyone feels,' pondered the doctor. 'Everyone hated that bust, but no one dared say it.'

If a state of Low Batt often proved fatal for the unfortunate individuals admitted to Mama Yemo, there was one public institution where the chronic condition held out potential risks for an entire city. On a hill looking out over Kinshasa is a one-storey building that briefly became the focus of local media attention a couple of years after Kabila's takeover.

A metal projectile, the newspapers reported, had ploughed into the wall of the establishment. As no one was hurt in the incident, debate focused on the puzzling question of where the missile had come from. True, a civil war was raging across the river in Brazzaville. But it seemed unlikely that a missile, however badly aimed, could stray as far inland as this. The other possibility was that the 'missile' was in fact a fragment from a small plane that had recently crashed near Ndjili airport. But, once again, it was difficult to work out how the debris could have ended up so far from any known flight path.

The matter would have remained of purely academic interest had it not been for one key point—the building happened to be a nuclear reactor, the first reactor ever built on the African continent. With its one megawatt capacity, it is dwarfed by the likes of Chernobyl's 1,000 megawatt installation. Nonetheless, if damaged, it could spew radioactivity for kilometres around, leaking contamination into the city's water supply. No wonder when I visited the site on the university campus, a white-coated technician, leaning over to inspect the punctured wall, was shaking his head: 'This is more than just worrying, it's a threat. If it had hit the centre it wouldn't have been funny at all.'

Few clues to the building's purpose are available on the meandering approach to the reactor. To get there, you must cross the district of Limete, stronghold of Etienne Tshisekedi. Once the champion of the opposition movement, he is now a stubborn old man who likes to sleep late and rarely strays beyond his own courtyard. The highway is a blaring ribbon of buses overloaded to the point where they develop a permanent slouch, honking Mercedes driven by army captains and taxi-buses with urchins dangling from the fenders, all maintaining top speed as they swerve around the potholes. At the end of the thoroughfare, you circle around the monument to Patrice Lumumba, erected by Mobutu to beatify the national hero he helped destroy. Kinshasa's equivalent of the Eiffel Tower, the monument was ultimately meant to hold a restaurant with panoramic views, but the cranes on its airy platform stopped moving long ago. Like Lumumba's nation-building project itself, it has never been completed; with every passing year, more panels and constituent parts go missing, cannibalised by pragmatic patriots.

At this point you turn inland, climbing through the district of Lemba and a series of haphazard markets redolent with the sharp stink of chicken droppings, where women sell bread baguettes from large metal basins, lorries load up young labourers, and girls sit patiently plaiting yellow, mauve and orange tresses into each other's hair. Eventually, the Ministry of Information and other skyscrapers become toy blocks shimmering in the haze of the valley and you reach the cool air and open green hills where the city finally begins to sputter to an end. This is the entrance to the University of Kinshasa and the sudden deluge of neat white shirts and young faces sheltering from the sun under coloured umbrellas is a reminder that this is a very young country.

Once on campus, there are no carefully monitored perimeter fences, guard dogs or electric warning systems. Only a small sign—one of those electrons-buzzing-around-an-atomic-core logos that once looked so modern and now seem so dated—alerts you to the presence of radioactive material. Behind the reactor's rusting gates, secured with a simple padlock, the courtyard resembles a wrecker's yard, littered with the rusting hulks of cars being tinkered with in the hope of eventual revival. The grounds are being put to the usual culinary use, with cassava bushes and papaya trees growing on either side of the main entrance.

The day I visited, the only formality involved signing a book held out by a man in a dingy sideroom. It was only later, when the head of the reactor mentioned that a crack team of gendarmes had been assigned to guard the building, that I realised I had already made contact with this elite unit. If at the Mama Yemo hospital, staff were all too bitterly aware of the extent to which a state on permanent Low Batt had abandoned them, Professor Felix Malu Wa Kalenga showed a blissful ignorance of that overwhelming reality. 'I have absolutely no worries about security,' he assured me, moments before casually mentioning that a fuel rod which had gone missing two decades earlier had recently, to the administration's astonishment, been unearthed by the Italian police, property of the Sicilian Mafia. 'I think one of the previous directors was a little careless with his keys,' he explained. 'He probably lent them to someone, not realising that the key to the reactor was on the same bunch.' Well, anyone can make mistakes.

Professor Malu was a gangly, spider-like man with long arms and legs, even longer fingers and a head of bristly greying hair. He wore a hearing aid, but it didn't seem to work. I asked questions and he gave long, detailed answers. Sometimes, for brief interludes, the two would coincide. But most of the time he answered questions I had not asked and I put questions he did not answer. His deafness made him seem evasive and unhelpful, but I suspect embarrassment at his affliction left this intelligent man constantly trying to second-guess conversations. The deafness also had its advantages for the veteran physicist. It reinforced a wall of inexplicable serenity he had built around himself in order to allow his pet project to remain in existence.

The missile impact and theft of the rod were not, after all, the only time the nuclear reactor had triggered a panic spreading well beyond the country's frontiers. Realising they faced defeat at the hands of Kabila's advancing forces, Mobutu's presidential guard had drawn up plans to blow the reactor up in 1997—plans that were luckily never put into action. Periodically, local politicians have warned of the risks of a landfall sweeping away the reactor. The university campus, like much of Kinshasa, is built on sandy soil disastrously prone to erosion, with entire hillsides regularly subsiding overnight. 'Obviously it's not a good thing having a land fall near a nuclear centre,' acknowledged the professor. 'But the last one was at least 100 metres from here. There was no real danger.'

Given that Kinshasa had lived through two rounds of looting, one military takeover and a major rebel attack in the last eight years alone, I suggested, wouldn't it be sensible to ask the International Atomic Energy Agency (IAEA) in Vienna to send officials to remove all radioactive material and close down the facility for good? 'Certainly not,' retorted Professor Malu. 'We have never had any real problems, although I do get the impression that the IAEA regards us with some suspicion because our country is in ruins.'

The fact that Kinshasa possesses a nuclear reactor at all, surely that most inappropriate of institutions for a country incapable of providing millions of its citizens with electricity or clean running water, is really a fluke of history. When I first heard of it, I automatically assumed it was a *folie de grandeur* on the part of Mobutu, one of the more perilous of his white elephants. But I was wrong. Kinshasa's reactor was a gift from God. It was the brainchild of Monsignor Luc Gillon, a ferociously energetic Belgian priest who had trained as a nuclear physicist, studied at Princeton, and in later life poured his energies into setting up Belgian Congo's first university. Like a colonial administrator who uses his years in the tropics as a chance to build up his butterfly collection, Mgr Gillon seized the opportunity to indulge in his hobby: nuclear research.

In the run-up to the Second World War, when the likes of Albert Einstein and Robert Oppenheimer were growing interested in nuclear fission, Belgian Congo was the world's biggest producer of uranium. Found in concentrations almost unheard of anywhere else in the world, the bright yellow substance was dug from the mines of Shinkolobwe, in the southern province of Katanga. But since the colonial authorities themselves had no interest in the metal—they sought the radium excavated alongside, used to treat cancer—the state-owned mining company Union Minière du Haut Katanga allowed a prescient director to ship three years' worth of uranium stocks to the United States, where it was on hand when work began on the Manhattan Project. The bombs the Enola Gay dropped on Hiroshima and Nagasaki in 1945 were made with Congolese uranium.

Under the secret supply deal signed with the US, Belgium agreed to sell Congo's uranium at a nominal price in return for American help funding its peacetime nuclear energy programme. For Mgr Gillon, it seemed only right that the colony that had provided the raw material that effectively ended the Second World War should benefit. The head of Belgium's atomic energy programme, perhaps understandably, disagreed. 'He believed an atomic reactor would serve no purpose in Congo and indicated that the American dollars on which I was counting for funding had all been spent,' Mgr Gillon recalled in his memoirs. But the priest, whose self-confidence comes across as bordering on bumptiousness, steamed forward regardless.

He set about buying a US-made, 50-kilowatt Training and Research Reactor for Isotope Production General Atomic (TRIGA) he had 'fallen in love with' at an exhibition in Geneva. Despite the fact that Leopoldville had been swept by violent riots just a few days before the Triga reactor arrived in Congo in early 1959, the question of whether this potentially hazardous invention would be an appropriate inheritance for an unstable government terrified of its own mutinous army does not appear to have troubled the well-intentioned Mgr Gillon. 'At the time people were talking about independence in 30 years' time,' shrugged Professor Malu. 'It wasn't envisaged.' Less than a year and a half later, Congo was on its own.

And so Congo became the first African member of the IAEA, an achievement that was a source of huge subsequent pride to Mobutu. In 1970, Gillon and Malu, his young sidekick, decided to upgrade the reactor to its present capacity. By this time, Shinkolobwe had closed and Congo could no longer provide the raw material required. It was forced to buy its own uranium back from the Americans. The humiliation still rankles, contributing to a profound feeling of grievance in Congo, where the secrecy surrounding the 1944 supply deal has left locals convinced Belgium made a killing on the uranium sales and that on this issue, as with so many others, their country was ripped off by a cynical West.

Professor Malu remains inordinately proud of having managed to carry out the upgrade. 'It was a very, very dangerous operation. One slip, and you could be irradiated. We had no help from anyone, we financed it ourselves and we did it all on our own.'

Why bother? I was tempted to ask. For the Triga reactor serves no conceivable practical purpose. It was never designed to provide electricity to a nation that in theory already had the hydroelectric capacity to export power to the region. Its *raison d'être* was purely educational: producing isotopes used in scientific experiments, such as irradiating seeds in the hope of producing disease-free varieties. Now even that abstruse function has fallen by the wayside, as producing the radioactive elements needed for such research costs more than buying them abroad, leaving the institution little more than a tempting, if ultimately unrewarding, target for the nimble-fingered. In one of Kinshasa's looting sprees—'après le deuxième pillage'—it was said, animals being used in experiments at the reactor were stolen and eaten, radioactive or not.

But this was to reckon without national *amour propre*. Mobutu had taken a huge interest in the nuclear reactor, making a point of attending all its special events and providing funding when emergencies cropped up. Kabila took a similarly benevolent approach, dispatching the thirty-man team of gendarmes to guard the installation. The new president was as unlikely as his predecessor to close down a facility he regarded, however ludicrously, as a symbol of prestige.

'In principle', said Professor Malu, the reactor was still switched on briefly once a week, to verify it was still functioning properly. But anything more ambitious was ruled out because of lack of funding. The general dearth of financing clearly weighed on his mind. 'The policemen guarding this facility earn $200–300 a month. That's ten times more than a university professor. C'est pas normal.' As the years went by, he acknowledged, the forty-year-old reactor was becoming increasingly difficult to maintain. The world of nuclear technology had moved on and Kinshasa's monitoring equipment had become obsolete, with worn parts impossible to replace. This forced the technicians to go in for a bit of 'bricolage' (do-it-yourself), he admitted, a comment that raised the hairs on the back of my neck.

Such tinkering might no longer suffice. Visitors to the reactor have reported that the water used to cool the rods is becoming grubby, a sign that it is impure. Left unchecked, the process could lead to the corrosion of the uranium rods and eventual contamination of the site. 'Something has to be done there,' sighed a spokesman for the IAEA. Responsible for the world's nuclear industry, the organisation was far from reassured by the findings of a safety inspection conducted at the Kabila government's request. 'There is a problem with the buildings' foundations and a general problem of a lack of infrastructural support from the government. They are aware this is not the best of situations and we are trying to help.'

Not so aware, apparently, as to recognise that Congo and nuclear energy should finally part company. Staff at the reactor approached the US with a bizarre deal in November 1999. In return for granting US experts permission to empty fuel from the area decommissioned during the famous upgrade, they wanted investment in a range of nuclear research projects. True to the unquenchable spirit of Monsignor Gillon, construction of a brand new nuclear reactor topped the list.

Some industry experts have speculated that if conditions at the Kinshasa reactor deteriorate beyond a certain point, Washington, increasingly nervous about the temptation offered to terrorists by such poorly policed nuclear rods, might feel obliged to stage a repeat of the operation in which nuclear fuel was removed from Vietnam before the country fell to the Vietcong.

Just how dangerous was the Kinshasa facility? I asked the IAEA. 'We did some back-of-an-envelope calculations when the rebels were advancing in 1997 and worked out that there could have been contamination if the reactor had been blown up, but that it would have been largely limited to the university,' an official told me. 'Certainly, from a safety standpoint, it lies at the lower end of the world league table. But while one can't be complacent, the scale of potential contamination is roughly 1,000 times less than Chernobyl, because the reactor is 1,000 times smaller.'

As I drove away from a battered reactor full of murky water, on a hill slowly sliding into the valley, run by physicists who had lost touch with reality, the figures somehow failed to reassure. Before I left, Professor Malu had gestured to a technician, who switched off what sounded like a car alarm, opened a locked room and emerged proudly holding a silvery metallic tube. The uranium rod was gleaming. To my untrained eye it looked pristine, free from any signs of corrosion. But in a nation on Low Batt, I wondered how long that could remain true.

Caesar and Brutus. As Patrice Lumumba (centre) forms his first government, Mobutu (far right, back row) waits in the wings. © *P. Rolda, Sygma*

Mobutu and Kennedy: A young president and an army chief reach an understanding during the Cold War years. © *Keystone-Sygma*

Keeping up appearances: a roadside hairdresser plies his trade outside the heavily-guarded Hotel Intercontinental. © *Reuters*

Pedestrians leap across a rubbish-filled ditch as Kin-la-Poubelle ('Kinshasa the rubbish dump') lives up to its nickname. © *Reuters*

A young practitioner of Article 15 touts a selection of hard-boiled eggs, cola-nuts, cigarettes and lollipops around bars and restaurants.
© *Yves Pitchen*

A sapeur struts his stuff in one of the capital's open-air nightclubs.

A student at the National Conservatory of Music practises on one of the last grand pianos in the city. © *Yves Pitchen*

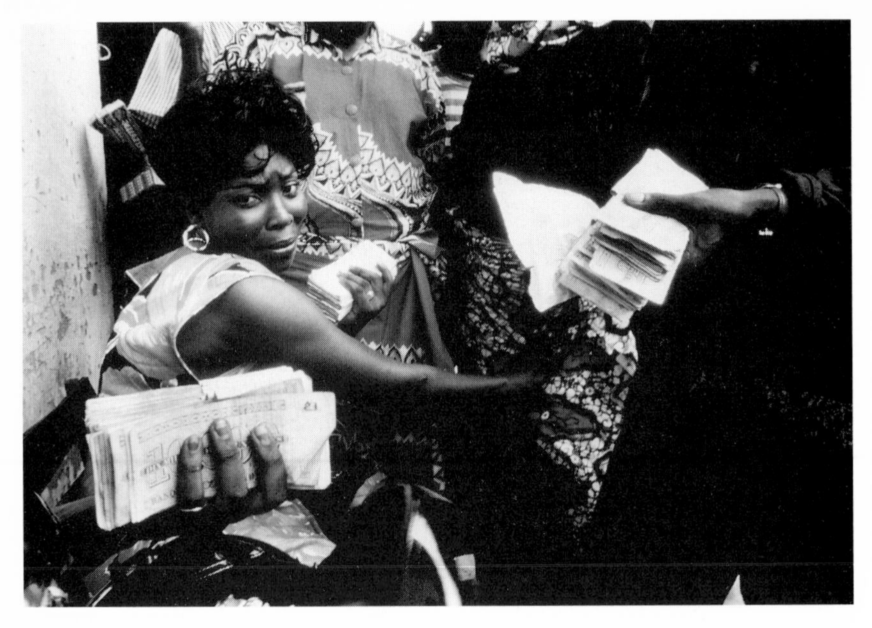

Not worth the paper it's printed on: banknotes change hands as the currency plummets on Kinshasa's 'Wall Street'. © *Malcom Linton, A L L Photographs*

A handicappé trader prepares for a river crossing to Brazzaville.

Kinshasa's street vendors used to sell model cars made of wire. When journalists flooded into the country to cover Mobutu's last days, the cars were replaced by TV cameras, snapped up by the media. © *Reuters*

Residents help themselves, looting local businesses after the retreating army pulls out of Goma in late 1996. © *Reuters*

A leopard in decline. © *Associated Press A F P*

CHAPTER SEVEN

Never naked

'A man from the Congo, living in Brussels, travelled regularly to London on Eurostar to collect housing benefit, an Old Bailey jury heard today.

'Ngolompati Moka, 33, who is a Belgian national, used fake tenancy agreements to persuade the boroughs of Hounslow and Haringey to pay him a total of £4,653.36, said the prosecution. The court was told that Moka, who was born in the Congo, used a series of identities to claim the cash. After he was arrested in a Hounslow JobCentre last August, police found a number of documents that incriminated him. These included bogus tenancy agreements, a Belgian ID card and receipts from Eurostar trains. 'These show he was making trips from Brussels to claim benefit in this country," said counsel.'

***Evening Standard,* 28 January, 1999**

'A mouse that goes hungry in a groundnut store has only itself to blame.'

Bas-Congo proverb

During the tumultuous post-independence years, when Congo looked to the outside world like an oversized fruitcake about to crumble into a hundred tempting morsels and an uncertain Mobutu was sizing up army morale, an empire briefly saw the light of day in the province of south Kasai, home of the Luba.

With the quiet blessing of the Belgian mining companies, more interested in guaranteed access to Congo's mineral wealth than issues of national sovereignty, the diamond province followed the example set by Katanga further south, and announced its independence. Its new emperor, Albert Kalonji, suddenly found himself swamped by returning Lubas fleeing an army which interpreted its orders to put down the secessionist revolts as a licence to massacre members of the ethnic group. Exasperated by constant requests for shelter, seeds, tools and money, he finally issued a statement telling the refugees to stop bothering the government with their problems, going so far, some say, as to write the principle into the empire's new constitution. 'Vous êtes chez vous, débrouillez-vous,' was the message—'This is your home, so fend for yourselves.'

Thus was born the infamous Article 15. 'Je me débrouille,' the Kasaians would say, with a knowing nudge and a wink, as they indulged in a bit of light diamond smuggling. 'Article 15,' Kalonji's officials would quip, with a philosophical shrug of the shoulders, citing the constitution to justify their demands for bribes. And later, long after the empire had faded into history, the principle received another top endorsement when Mobutu, addressing a ruling party conference, acknowledged that it was acceptable to 'steal a little', as long as the theft remained within limits.

By the 1990s, Article 15 was the sardonic thread running through the fabric of Zairean society, the *raison d'être* of a leader, a government, an entire regime. Prime ministers came and went, each of them doling out civil service jobs for the boys. There were official drivers for ministries without cars, switchboard operators for departments without phones, secretaries without typewriters. They were paid an average of $6 a month but hung on nonetheless in the hope that one day the economy would revive and they would get what they were owed. In a world of fantasy wages, knowing how to 'se débrouiller'—that untranslatable French concept meaning to fend for oneself, to cope against all odds, to manage somehow—had become something approaching an art form.

For public servants, juggling two jobs—the official one that involved sitting in a dimly lit office reading the newspapers and the real one that started at noon and, hopefully, brought in some real money—became the norm. The skill was finding a Darwinian niche in the ecosystem, the tiny competitive edge that meant one had something to sell, a means of survival in a ruthless world.

For teachers it was the pass marks they could dole out to ambitious pupils in exchange for groceries. For soldiers who rented themselves out to private businessmen or restaurants as bodyguards, it was the promise of security. For diplomats in embassies abroad, it was access to duty-free goods. For bureaucrats, it meant touting their ability to delay or expedite crucial paperwork and influence with superiors. As Mobutu himself once remarked in yet another of his surprisingly frank conference speeches: 'Everything is for sale and everything can be bought in this country. And in this trade, the slightest access to power constitutes a veritable instrument of exchange.'

But in a shrinking economy, most Zaireans never succeeded in entering that world of pensions and salaries, however many months in arrears. In 1955, nearly 40 per cent of the active population worked in the formal sector. By the 1990s, this had shrunk to 5 per cent and the official figures for per capita income had fallen to a laughable—and obviously impossible—$120 a year. Zaire's real economy had dropped off the map. The vast majority of Zaireans were living off the ubiquitous vegetable plots and their wits, buying and selling, smuggling and haggling, hustling and rustling.

Once, sitting in a working man's restaurant behind Kinshasa's main post office, a monolith sprouting an impressive array of obsolete antennae and dishes, I logged the stream of street sellers who entered, trawled and departed. It was a walking supermarket whose players ranged from established traders, with cigarettes and sweets displayed in usherette-style boxes, to those who seemed to be operating on an ad-hoc basis: one second-hand shirt slung over an arm, a single belt dangling from a finger, a couple of well-fondled chocolate bars whipped out of a pocket.

In the space of forty-five minutes, as I worked my way through a steaming plate of rice and beans, I was offered the following items without straying from my seat: cigarettes, chewing gum, hard-boiled eggs, cola nuts, spice sachets and carrots (all from a medicinal box aimed at those plagued by bad breath or sore throats), French perfume (two tatty boxes, clearly fake), plastic briefcases and plastic sandals (range of), a shoe polish (a small boy knocking his brush against a stool to attract attention), men's trousers, transistor radios (choice of two models), a display of tinny-looking watches and sunglasses, ginger powders, a couple of sports shirts, cheap nylon ties, disposable razors, men's briefs (packet of three), men's shirts, paper tissues, roasted peanuts (in the sachet), grilled prawns (on wooden skewers), socks (variety of colours). The traders patiently allowed their goods to be examined and commented on by the sceptical but not unfriendly diners, then moved tirelessly on. It was like watching predators on the savannah as they prowled the long grasses and scoured the horizon, searching relentlessly for a kill.

The variety of forms of Article 15 adopted in Kinshasa never ceased to amaze. In the markets, there were the wood-carvers who had perfected the art of making masks, stools and fetishes look like weathered antiques, the starting point of a chain that ended in

European galleries, where the supposed *objets d'art* sold for hundreds of times their original price. On street corners, boys working in cahoots with Post Office staff sold stolen copies of *Time* and *Newsweek* ordered by the few expatriates still foolish enough to expect to receive such tempting mail. At rush hour these youths might metamorphose into the bully-boys who lined the bus routes and would, for a price, shove their way onto a packed minibus, clearing a seat for commuters who didn't want to crease their clothes. Their brothers operated in the money-changing districts of the Cité as 'chargeurs' whose job was to rush any approaching car and lure its passengers to a foreign exchange bureau, benevolently ignored by the traffic police who themselves sold Zairean driving licences on request. And then there were noble, solo efforts, such as the man who stood in the middle of Avenue Colonel Lukusa day after day, gesturing melodramatically at the hole in the road he had filled with sand for the benefit of passing cars and demanding, in increasingly outraged tones, to be paid for his efforts.

For sheer weirdness, though, it was hard to rival what went on at Ngobila Beach, the port where the lumbering river ferries landed and my own first point of contact with Zaire. This was the main link between Kinshasa and Brazzaville, the capital on the other side of the swirling brown waters, and like all border crossings it provided a million ripe opportunities for Article 15.

The fact that I was proposing to have a drink with my two unconventional companions so amazed the owner of the Mona Lisa, the café around the corner from the beach, he was initially lost for words. After twenty stunned minutes, however, the agony had got too much to bear and he ushered us indoors, off his exposed front terrace. 'Do you KNOW these men?' he asked incredulously, upper lip visibly curling in distaste. 'Please keep it brief,' he begged. 'I don't like these people.' Our drinks, predictably enough, never appeared.

To be fair, we made for a fairly unconventional party. Neither of my guests was particularly clean, due to what was clearly regular contact with the ground. As a group, we probably smelled fairly rank and between the lot of us we could only boast three sound legs—two of mine and one belonging to Ntambwe Mpanya, president of the Ngobila Beach Handicapped Mutual Benefit Society. He was a broad-chested fellow, with a round, extraordinarily expressive face. But a shrivelled left leg had put him on a permanent kilter. He walked with an ugly, wildly swooping motion, plunging earthwards with every second step, only to be saved from impact as his sound leg kicked in.

General secretary Zege Osenge was even worse off. At first glance he appeared to have no legs at all. Closer inspection revealed two useless appendages incapable of bearing weight. Sitting in a plastic café chair, he seemed normal, a fine figure of a man, in fact. But as soon as he left the chair a terrible transformation took place: plummeting to hip height, he became a vision of horror, the kind of logic-defying deformity that rises gibbering and scrabbling from the depths of the subconscious at night. The agility with which his truncated form hopped into cars, negotiated doors and scampered along pavements was so unexpected, it made one catch one's breath in something more than mere disbelief.

It is hard to think of a place where life is harder for the disabled than Africa, a world away from the linguistic euphemisms, collective guilt and uneasy piety of the West. Handicapped children—for the most part, victims of polio outbreaks their parents could not afford to vaccinate them against—are often not sent to school. As adults they struggle to find work, get overlooked in hospitals and are dismissed as unsuitable candidates for marriage. The instinctive shudder of horror, the automatic shrinking from contact, are barely masked by the able-bodied. It is an approach whose only merit is its complete lack of hypocrisy.

In Congo, poverty and desperation have increased the pressure on families tempted to throw these financial burdens out of the house, to join the blind men and sun-blistered albinos who gather at crossroads to beg from passing cars. There they swell the misshapen ranks of the gangs who effectively run a Mafia-style system, organisations based on the realisation that while one man with no legs can achieve little, a score of them demanding action can command considerable attention. Touring local stalls and shops in force, they threaten to smash windows and block entrances unless shopkeepers pay regular tribute. The sight of a crowd of furious cripples gathered outside his premises on their wheeled palettes and tricycles is usually enough to persuade a shopkeeper to pay up. But when roused, the paraplegics have been known to systematically hunt down their enemies in packs. To the physical fear of the able-bodied, appalled by such naked aggression, is added another concern: the suspicion that the handicapped possess evil powers, supernatural compensation perhaps for the rotten hand they have been dealt by fate.

It was specifically to avoid resorting to such gangster tactics, explained Ntambwe, that the Mutual Benefit Society was created. With 213 members, both men and women, it now effectively dominated commercial trade from Ngobila Beach and was the only paraplegic association, he claimed, that managed to make a living. 'Article 15, that's us. We were determined not to be beggars. Because no one out there would give us any work, we decided, as intellectuals, that we had to go out and create it for ourselves.' Each morning, scores of paraplegics gather at the warehouses in the narrow streets off the port. Sitting on their hand-pedalled tricycles, they wait anxiously while their young assistants load up the specially designed compartments at the back with goods. When the tricycles can take no more, they head for the riverfront, drivers straining at the handles and sweating helpers pushing from behind. Each evening the same bizarre procession wends its way back from the port, equally heavily laden, but with a different range of goods.

When it comes to competitive advantages, few can be more fragile than this. The society owes its commercial viability to a quirk of law which allows cripples to travel on the ferries at a discount. The handicapped travellers could accordingly afford to set slightly lower prices for their goods and thus became the favourite go-betweens for the ferocious 'Soeurs ya Poids' (Heavyweight Sisters), the buxom merchant women who sell in the sprawling markets on either side of the river. The paraplegics also enjoy another advantage. Because officials shrink from touching them, terrified of the cripple's curse, they pass through the frontier with a minimum of inspection and cursory customs charges. They are therefore perfect conduits, say Kinshasa's residents, for drugs, foreign currency and any other small, precious items an exporter prefers not to declare.

But the real skill is to exploit the temporary, short-term scarcities that develop in each capital city, whether rice, milk, flour, sugar or margarine. 'We bring from Kinshasa what is missing in Brazzaville and what is missing in Kinshasa we bring from Brazzaville,' explained Ntambwe. 'Sometimes, if we get it wrong, we have to bring our stuff all the way back and the day has been lost.'

Not far from where we sat, a paraplegic was busy capitalising on the latest twist in market forces. Helped by friends, he was struggling to balance a pair of giant jerry cans filled with petrol onto the back of his tricycle. At that moment, the militia fighting in Brazzaville meant fuel there was scarce. Kinshasa's petrol, itself in short supply, should sell for a high price over the river, high enough, in any case, to justify this polio victim running the risk of becoming a tricycling firebomb if a cigarette spark went astray. 'Thanks to the war, I should be able to sell the petrol on the other side for twice the price,' he said. 'Then I'll bring milk back in the same jerry cans.'

Key to the trade were the hand-pedalled tricyles. Manufactured in Kinshasa to precise instructions from the paraplegics themselves, they came in a range of shapes and styles. A light, mobile one served for rushing around town. Those used for trade were heavier and more bulky, equipped with metal compartments on the back. And then there was the handicapped equivalent of the long-haul vehicle—tricycles with storage tanks so large they could not be manoeuvred without two helpers heaving from behind.

These young boys, the 'aides-handicapés' were also critical to commercial success. 'It's very important to be able to trust your assistant,' explained Zege. 'The handicapped person can't go and buy the rice or flour at the market himself, so he has to be sure that the assistant isn't cheating him. Which is why the assistant will usually be a member of the family.' For both players, it was a job that demanded aggression and a high level of immunity towards the endless 'tracasseries' (hassles), the shouting, the wheedling negotiations, the temper tantrums that accompany any frontier crossing in central Africa. 'Policemen are the Number One enemy of the handicapped,' complained Ntambwe, his features disfigured by a terrible scowl. 'They treat us worse than rabid dogs. For the officials at Ngobila we are not human. At the other end, in Brazzaville, they behave just as badly. The UN talks about the rights of the handicapped, but here we have none. It is an African disease.'

Try as he might, the president made an unconvincing victim. There was something about his cunning face that revealed a survivor; a thuggish belligerence born of necessity and laid down over the years, layer after layer, a carapace against a hostile world. And indeed, the game had clearly been worth the candle for its leading player. Struck down by polio at the age of one, Ntambwe was so determined to get on in life he used to follow friends to primary school in defiance of his parents' wishes until they reluctantly agreed to enrol him. He got some education, but never as much as he would have liked. Now he was hailed respectfully as 'president' wherever he went, owned a motorised Vespa worth $800, and managed to support a wife and eight children. He nursed expansionist plans—he wanted to buy a second Vespa for the society's communal use, and dreamed of running training courses in shoe-making and poultry farming for its members.

His secretary, who was paralysed as a result of a bungled injection, must have hoped for better than this when he was a young man studying law at Kinshasa University. He blamed the premature collapse of his academic career on a rector who took against his unsightly student and had him expelled from campus. But in managing to support a wife and three children, he was doing as well, if not better, as many unemployed former students who had once believed their status as 'intellectuals' would automatically land them a white-collar job.

The fact that both men were earning their living, they said, had brought about a miraculous change in attitude by initially hostile in-laws, who had come to realise the husbands they feared would shame the family were bringing home a decent wage. Neither of them fooled themselves that the sudden affection went very deep. 'If you show signs of succeeding in life then you'll find the in-laws coming after you,' said Zege. 'Recently we had to buy the coffin of a society member who had been beaten up and killed by a porter. His relatives made themselves scarce as soon as there was no more money coming in. When you're handicapped, you can only ever rely on another handicapped person.'

If the Mutual Benefit Society was an example of Article 15 at its most inventive, it also highlighted how easily those survival techniques could be crushed, the competitive edges eliminated in a moment by a minute change in the law.

I lost sight of the two for a couple of weeks, but kept cruising the harbour in search of them. On my last day, by chance, I spotted the president on his beloved Vespa, returning from another round trip to Brazzaville. François, with his usual blunt driving style, decided to attract the president's attention by mounting the kerb in front of him. I noticed with amusement Ntambwe's instinctive snarl of aggression and furious yelp at yet another insult directed at him from the able-bodied universe, before he realised he was being hailed by a friend and the scowl turned into a smile.

Things were going badly for him personally and the society in general, he complained. The Vespa had been in a minor accident, emerging with a dented bumper. Simultaneously, the Kabila government, so desperate for money it had taken to taxing the money-changers and hairdressers who worked by the roadside edge, was refusing to recognise the special privileges handicapped travellers traditionally enjoyed. Touring the port, the finance minister had recently made it clear he regarded the society's members as little better than smugglers working in collusion with the Heavyweight Sisters to cheat the government. 'The state has never done anything for us. We have no pension, no social security. If my child gets sick and needs hospital treatment, that's it. And now they tell us we must pay just like everyone else. We don't know where to turn.' As a result of the government's new approach on customs and ticket charges, he said, the paraplegics' profits had already fallen to a quarter of their usual level. 'They're squeezing us hard,' the president muttered, with a shake of the head. 'We must get these measures changed, because we can't go on like this.'

The mood was gloomy as I accompanied the two executives down to the waterside to watch the last berthing of the day. The police and paraplegics had already assumed their respective positions on the quayside. As ferry grated against dock, one stout officer began flailing around with a long black whip, determined to stem the flood of paraplegics attempting to board for free. Shouts escalated, the whip thudded against wood, the security forces appeared to be winning the day. Then I spotted one of the Mutual Benefit Society's members. So stunted he seemed no more than a head attached to a pair of sinewy arms, he had exploited his size to quietly squeeze through a gap in the wire netting separating us from the water. Hand over hand, he swung his body along the underside of the ramp being blocked by the police. Clambering aboard behind them, he turned, grinned triumphantly and pointed at the adversaries he had just outsmarted, mouthing something. I could not understand the Lingala, but somehow I knew it was very rude.

'Never naked,' shouted Charles, pounding the steering wheel for added emphasis, as though any were needed. 'No, no, Madame, never naked!'

As we negotiated the centre of town in his gleaming four-wheel drive, Charles was running through the finer points of his religion. For Charles was a Kimbanguist, and the Kimbanguist Church, it turned out, had a bit of a thing about drinking, dancing, polygamy and personal nudity. A true believer, he explained at the top of his voice, would never undress completely, retaining his underwear even in the shower, bath and bed. Christ himself, he pointed out, had retained his briefs during baptism. In the presence of God, it was vital to be decently dressed, and God was present at all times, after all.

For the paraplegics of Ngobila Beach, facing a future without social security, Article 15 was something to be openly embraced, acknowledged without embarrassment. But even in the despair of the modern-day Congo there were those, I discovered, who felt obliged to dress their coping mechanisms in more respectable clothes.

There was something of the rubber ball about Charles. A roly-poly figure of a man who seemed to have been born without a neck or any other superfluous links between his various body parts, he was blessed with unsquashable cheerfulness. Swat him away, one felt, and his tubby form would merely bounce off the nearest wall and come hurtling back, smile intact, perky as ever.

Two-way conversation was an alien concept. Charles was into pronouncements: emphasised by a stubby finger, delivered with all the force and rhythm of the amateur preacher he was, they brooked no dissent. Beaming with enthusiasm, he would round them off with an emphatic 'Merci', signalling his point had been not only made, but proved. His colleagues had dubbed him respectfully 'Papa Pasteur'. He had a collection of Kimbanguist religious books, attended church whenever he got the chance and would clearly have liked to be at worship every day of the week. Proudly, he showed off the little diamond-shaped badge—a portrait of the church's spiritual head—he always wore pinned to the natty waistcoats he favoured, and fished a plastic eye-drop bottle out of his rucksack. It was full, he explained, of Holy Water from the village of N'kamba. Birthplace of the prophet who founded the church, N'kamba had become a place of pilgrimage for believers, the Kimbanguist equivalent of Jerusalem. 'Every morning I pray and swallow a couple of drops of this. It keeps me blessed all day.'

Yet in this picture of religious devotion one note jarred: the slight matter of his job, which, it seemed to me, might present a committed Christian with a few pangs of conscience. For Charles was what was known as a 'protocol'. Employed by one of the few multinationals still operating in Kinshasa, his duty for the last thirteen years had been to usher employees in and out of Ndjili airport with a minimum of hassle. Essentially, that made him a professional payer of bribes. As a committed Christian, did this not present him with something of a moral dilemma?

No job goes deeper to the heart of Article 15 than the euphemistically dubbed 'protocol'. In most airports of the world, all but the sick, very old and very young can be trusted to show their tickets, check in their luggage and pass through customs on their own. The fact that in Congo an entire profession has sprung up to deal with these simple procedures is a tribute to the sheer inventiveness of the country's officialdom.

Under Mobutu, the airport was transformed into something far more challenging than a place where you merely boarded a plane. Every check represented an opportunity for poorly paid officials to exchange their co-operation and compliance for a 'little present'. And so the checks multiplied—at one stage there were seven separate services with the right to examine your papers—as members of the rival security forces, with friends, relatives and hangers-on attached, moved into Ndjili's peeling cupola with their snarling police dogs, metal barriers and rubber batons. The building became more than an airport, it was transformed into an intellectual contest, a real-life computer game full of hidden traps and sudden obstacles, where the punishments ranged from public humiliation to arrest and the prize—that longed-for seat on the next plane out—came with a highly variable price tag.

The state of the airport mirrored the state of the body politic, offering new arrivals an accurate thermometer of the situation outside its walls. Every now and then, a new government flush with confidence would try to clean things up. A general would be given responsibility for the airport, the number of security services would be slashed and the human flotsam and jetsam expelled. But like hyenas lured by the smell of blood, the predators gradually crept back. As the new administrators registered the impossibility of reform, they would join the pickpockets, shoe-shine boys and beggars in the competition to see who could milk passengers most effectively. If you passed through relatively smoothly, you could count on finding some semblance of order outside. Emerge in a state of hysteria, several hundred dollars the lighter, and you knew things had gone bad again.

I never flew into Ndjili without feeling slightly nauseous. Although I knew it was coming, my stomach would lurch when an official at the door seized my passport and plunged with it into the bedlam of the interior, a move one felt was carefully calculated to put the nervous passenger off-balance. Travellers who could not afford a protocol developed their own survival techniques. My own was to keep a layer of dirty clothes at the top of my luggage, discouraging further exploration; to travel with the smallest of bags and always keep $20 notes ready folded in my pocket, to be smoothly pressed into a palm during the simulated friendly handshake that closed the ordeal. Getting the timing of the payout right was crucial. Volunteer the first bribe—intended to cover all officialdom in your ambit—too early and the friendly official who had earlier promised to 'arrange everything' would suddenly disappear, to be replaced by a fresh set of uniforms, all claiming to deserve a piece of the action. Best was after you had entered the taxi and the driver had his foot on the accelerator, but before one of the soldiers circling like sharks in the parking lot noticed money was changing hands. It wasn't always easy.

In this jungle, the experts were the protocols. Because they were regulars, they could trade favours, knew who mattered and the real price for each service. If they were good at their job, they took on their shoulders all the sordid demands of corruption, the dirty deals and petty bargaining, allowing their charges—the Big Vegetables and expatriate businessmen—to sit in the air-conditioned VIP lounge with empty hands and quiet minds. 'My responsibility is to make sure no one bothers my whites,' explained Charles, with possessive solicitude.

But it was not with this in mind, presumably, that Simon Kimbangu had taken on the might of the Belgian administration. For the founder of the Kimbanguist church was very much a rebel, deemed such a threat the colonial authorities imprisoned him for thirty years. Above all, he had very little time for the whites Charles spent his working day protecting from the more unsavoury aspects of Congolese existence. Which was why I found Charles's religious convictions, which he happily discoursed upon as we headed for the largest Kimbanguist church in Kinshasa, so intriguing.

Watching a religion and its myths while they are actually in the making is a curious sensation. The story of Kimbangu's life is full of parallels to the story of Christ and is told in much the same terms: his miracles, his twelve apostles, his initial reluctance to accept his divine destiny, his eventual martyrdom and posthumous apparitions. But the stories Charles was recounting had not been slowly crafted over 2,000 years, taking on an allegorical quality through the passage of time. Kimbangu died in 1951, within the memory of many living Zaireans, and the tale is dotted with disconcerting references to such modern inventions as trains, revolvers, ferries and motorcades. This is the Christian myth, replayed in the twentieth century.

Blurred photographs of Kimbangu exist. They show a man in a prison tunic, surprisingly stout despite his repeated bouts of fasting, frowning into the bright African sun. His scowl is deep and there is no discernible hint of charisma. But if you believe, as the church followers do, that Simon Kimbangu is not just a prophet sent to spread the word but the Holy Spirit itself, then here is that mysterious Christian entity, captured incarnate on film for the very first time.

He was born in 1887 in N'kamba, a village west of Kinshasa, where the river breaks up into a series of cataracts on its descent to sea level. The region of Bas Congo has always been imbued with a strong spiritual streak, thanks to its proximity to the coast, which meant it was the area the Portuguese missionaries first came into contact with. According to the legend, Kimbangu began receiving messages from God in his teens. By the time he married he knew a special destiny awaited him. In the 1920s, word spread that he was healing the sick, raising the dead and restoring the sight of blind people in the name of Jesus Christ. At first, the white missionaries had welcomed the efforts of a man who could convert local villagers to the Christian faith with such ease. But as it became clear that Kimbangu was challenging what he regarded as an attempt to establish a white monopoly on the Christian religion, a message bound to find a ready audience in a population smarting at its colonial subjugation, the Belgian authorities realised they were facing a potentially dangerous rebellion.

As Kimbangu's message that a black Messiah was coming who would expel the whites gathered pace, workers abandoned factories, refused to pay taxes and challenged the rules of forced labour. They flocked to N'Kamba in their thousands. 'The whites will become black and the blacks will become white,' Kimbangu preached, spreading a gospel in which anti-colonialism, black pride and personal salvation were inseparably interwoven.

Alarmed by a movement spreading faster than they could control, the authorities arrested him after a long game of hide and seek. At the end of a trial regarded by Kimbanguists as the equivalent of Jesus Christ's appearance before Pontius Pilate, Kimbangu was condemned to death after being found guilty of threatening state security. The sentence was later commuted to life imprisonment and Kimbangu was sent to what was then Elizabethville and is now the southern city of Lubumbashi.

While he served time, the Belgians tried unsuccessfully to wipe out the phenomenon, outlawing the movement, arresting its followers and deporting thousands of Kimbanguist families, a tactic that merely served to spread the word beyond Bas-Congo and further across the country. After thirty years in jail, where punishments included plunging the prisoner into saltwater after a beating, Kimbangu expired. While the colonial authorities ascribed his death to dysentery, believers say it was a death foreseen by the prophet himself. It was only when independence loomed on the horizon in December 1959, that the Belgians, accepting the unquenchable popularity of the movement, agreed to decriminalise Kimbanguism.

In 1960, the prophet's body, which witnesses said showed no signs of corruption after nine years, was exhumed and taken to N'kamba for reburial, receiving full military honours on the way from the army that had once hunted him down.

Kimbanguism is the antithesis of humility. Despite the puzzling absence of references to him in the Scripture, to millions of followers in Congo and neighbouring African countries, Kimbangu ranks alongside the Son and the Father as a constituent part of the Holy Trinity, a claim that must surely make the Vatican cringe. And what Jesus could do, worshippers like Charles make clear, Kimbangu did better, even if, to the outsider, the tales possess a certain wackiness that suggests symbolic reinterpretation needs to run its course if the story is ever to rival the New Testament and reach a wider audience.

Kimbangu, we are told, once foiled a plot to kill him with a slice of poisoned chikwange, that dietary staple, the local equivalent of trying to poison someone with a chip butty. Promenading himself on the River Congo, he not only walked on water but actually went one better. Soap and towel made a miraculous appearance, and he washed and dried his hands. When the colonial authorities were about to take him to prison, the train engine stalled for a symbolic three days while he said goodbye to his children. ('Three days, Madam,' exclaimed Charles, jabbing his finger at me. 'Three days! And no one could understand why.') And when the colonial authorities performed an autopsy on his body they found internal organs such as intestines, liver and lungs were missing, which makes the paunch captured by the camera all the more puzzling.

Such eccentricities did not stop Mobutu recognising a useful icon for a young nation when he saw it, and he sought to appropriate Kimbangu in much the same way that he had appropriated Lumumba. The first stone of the church's administrative centre in Kinshasa was laid by Mobutu. He also maintained cordial relations with the church's leadership, whose members were believed to enjoy special privileges as a result. In the eyes of some Congolese, the Kimbanguist clerics' disinclination to join the Catholic and Protestant churches when they started putting pressure on Mobutu to institute real democratic reform tainted the church.

But such charges left Charles indifferent. 'No man is a prophet in his own country,' he said with a shrug as we entered the gates of the Kimbanguist compound. 'People come all the way from Angola and Zambia to meet our spiritual head. But there are people in N'kamba village itself who don't believe in Kimbangu. Can you believe it?' Indeed, a half-hearted Kimbanguist, a Kimbanguist who didn't attend church, seemed hard to find. Maybe because of its clever interweaving of spirituality with the touchy issues of race and power, Kimbanguism seemed to have a knack for tapping a well of fanatical fervour in a population that had supped deep on the cup of humiliation. 'God is black,' explained Charles. 'The Pope said so when he visited Lagos in the 1960s. He said "God is black and he can be found in Africa". Well, we know who he was talking about. He was talking about us.'

As we wandered across the grounds, climbing through the scaffolding of what would eventually be a massive 4,000-seat Kimbanguist theatre hall, matching an equally massive hospitality suite next door—as genuine an example of African presidential kitsch, with its ice-cold air-conditioning and rows of chintz sofas, as any I had seen in my time in Kinshasa—I kept trying to bring the conversation round to the tricky subject of Charles's job. It was like trying to peel a mango with a knife and fork—at each attempt the subject skidded away as Charles skilfully rerouted the conversation back to his true interest: religion.

Things really were much better at the airport, he insisted when pressed, now that the Kabila administration had taken over. The airport had been cleaned up, the scum had been thrown out of the building and anyone paying a bribe risked arrest by the undercover agents working at Ndjili. As long as he was wearing his badge proving he was one of the several hundred accredited protocols—he proudly brandished his card—then there really was no problem. 'The Congolese want to change mentally. They are sick of what happened in the past. God is watching us and saying “the moment has arrived”. The black man must learn to know himself.’

Well, yes, he said, it was true that there were still five separate security services operating in the airport. And yes, their officials did still often ask for money, just to buy a ‘Sucré’—a sugared drink. ‘But you would have to have a heart of stone to refuse when a man who tells you his children can’t afford to go to school asks for just one Coke. As a Christian, I can’t see suffering and not be moved by it.’ Yes, he acknowledged, entry to the air-conditioned VIP lounge still required a ‘little present’. How much? Oh, quite a hefty present, $30 dollars or so. And yes, it was true that the expatriates who passed through his hands wanted the present to be paid rather than risk having their luggage inspected by officials. The most obvious question hung unspoken in the air between us: if Article 15 had really been banished from Ndjili airport, why should anyone need a protocol? If the new system was so clean, why wasn’t Charles redundant?

There was a pause while Charles indulged in a little uncharacteristic squirming. ‘Look, there are things I have to do for my job which, as a Christian, I clearly shouldn’t be doing. But our spiritual leader has told us, “your work is your mother and your work is your father”. So if it is done in the context of your job, there’s no problem.’

Once again, the stout prophet of N’Kamba was echoing Christ. ‘Render unto Caesar that which is Caesar’s,’ the Messiah had said, authorising his followers to pay Roman taxes and work alongside a terrestrial regime whose values they theoretically abhorred. The Kimbanguist version of that instruction for the twentieth century was as pragmatic as the prophet’s miracles and as hard-headed as Mobutu’s own advice: ‘Article 15 is acceptable, as long as it’s done for professional reasons.’

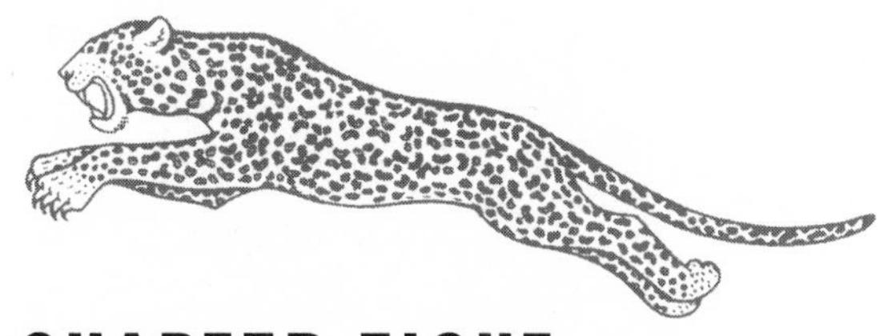

CHAPTER EIGHT

The importance of being elegant

'It seems that our country has been
Abandoned to its sad fate
Why did we fight against the white man's rule?
Did we shed our blood for independence
To listen to the sterile quarrels of our new masters
Fighting solely for their political privileges?
The country is in ruins.
What a humiliation before the world!
A country so rich, with leaders so careless of its
future
The time has come
Kasavubu, Lumumba, Bolikango, Old Tshombe
Why have you turned your backs on your country?'

—Song by
Tabu Ley Rochereau

If Article 15 was the pragmatist's reaction to privation, not everyone chose to follow the route of scrimping and scraping, making do and making it up as you went along. For some, mere survival was too small-minded an ambition. Their way of dealing with desperation was the path of sublimation, escaping into parallel intellectual or spiritual worlds where the rules were benign and self-fulfilment, grandeur, dignity—those qualities so missing in their daily lives—finally became possible.

You could hardly take two steps in Congo without stumbling upon a meeting of Seventh Day Adventists, a Moonie reunion, Jehovah's Witness get-together or a gathering by one of the US-imported fundamentalist sects that blossomed into new life in Congo's fevered climate. But the secular forms sublimation could take were more interesting. One case, in particular, first intrigued and then thoroughly alarmed both the Mobutu and Kabila administrations.

It came to public attention with a small march on Kinshasa's broadcasting centre, about a year after Mobutu's overthrow. The thirty or so members of the procession came from Makala, the working-class district in which Kinshasa's notorious prison was located, and they wanted national radio to transmit their simple message: President Laurent Kabila was to step down, return to his native Shaba province, and make way for the only man with the moral authority to rule: Mizele the First, the King of Kongo.

The security services broke up the procession without difficulty, but their antennae were now emitting loud bleeping noises—just what *was* going on in Makala? There was talk of some kind of royal court operating out of a modest local home. An army unit was sent to explore and took the precaution of staging a dawn raid. Not enough of a precaution, it transpired, for its soldiers were met with a volley of shots from those inside, who appeared to be sitting on a sizeable weapons cache. In the resulting day-long firefight between members of the 'royal court' and the security forces, which sent thousands of Makala residents running from the district in terror, at least eight people, including two 'kadogos'—the young boys making up the mass of Kabila's army—were killed.

When peace was finally restored a bizarre picture emerged. For years rather than months, the security services discovered, a little old man who made up in charisma for what he lacked in coherence, had been beavering away in the small house in Makala, busily setting up the structures of a state within a state. His plan was to recreate the ancient Kongo nation, that sophisticated kingdom discovered by the Portuguese explorers of the fifteenth century and destroyed by the slave trade and civil strife, with him as its first monarch.

Bernard Mizele Nsemi's efforts were already known to the local authorities but had been brushed aside as the ravings of a lunatic. But if the authorities had refused to take him seriously, his followers had not been so dismissive. They numbered in the thousands and came from all ages and social classes. They held ID cards issued in the name of the Kingdom. Fathers signed up their children, uncles their nephews, often, it later emerged, without the knowledge or approval of the individuals concerned. They even paid minimal levels of tax, a gesture of recognition and respect most Congolese citizens had long ago stopped according the state.

With the Congolese delight in hierarchy and status, the constituent parts of a nation had been identified and the top jobs distributed, from the Royal Prime Minister to the Interior Minister, Central Bank Governor to the Mayor of the port of Matadi. There were even bizarre echoes of Mobutu's divide-and-rule tactics in the multiplicity of security forces created and army chiefs nominated. In the short time that the royal court had existed, one felt sure that the Head of the mythical Royal Army had already been at loggerheads with the Chief of the Royal Police Force, who no doubt was bickering with the General Army Commander, who felt he couldn't trust the Royal Police Commander and suspected the Head of the Mixed Kongolese Army of plotting a *coup*.

When the military court convened on 9 July 1998 to try the 118 suspects on charges of murder, criminal association and plotting to overthrow an established government, the crowd that gathered to watch proceedings was so large the overwhelmed magistrate at one point thought of holding the session in camera. If idle curiosity played its part, another factor helped explain such keen attendance. Here was a group, the onlookers recognised, which had gone from complaining about the hardship of daily existence and dreaming vaguely of a better life—the bread-and-butter of every Congolese citizen's existence—to trying to bring about that future with the directest of methods.

Like David Koresh's Branch Davidians, they had at some point crossed the line dividing harmless fantasy from violent action. As in Waco, the authorities were about to signal just what they thought of such alternative worlds by crushing them. The government had moved from poo-poohing King Mizele as a comic irrelevance to attributing the entire movement to a European-backed destabilisation attempt, before concluding that this was in fact a home-grown rebellion, similar in nature to their own recent uprising, endowed with dangerous mystical overtones and fully capable of capturing the imagination of at least part of the population.

The trial, which received saturation coverage in the press, was a major organisational challenge. Mizele and the other suspects either refused or could not afford to defend themselves, so a team of eleven defence lawyers was appointed by the state. When the suspects were assembled in one spot they looked, in their glossy blue shirts—emblazoned with a giant 'P' for 'prisoner'—rather like an oversized football team limbering up for a match. But the crowd did not come to see the nobles of the royal court. They came to size up King Mizele the First, the man responsible for all the trouble.

He looked mild-mannered enough, a grey-haired old man, everyone's favourite uncle. But he had the single-mindedness of visionaries and madmen, making up in certainty what he lacked in royal blood. King Mizele did not base his claim on descent from any traditional chief. 'He told the court he was inspired by God,' said his attorney. 'He said he was in touch with our ancestors, including Simon Kimbangu and Joseph Kasavubu, Congo's first president.' Challenged by a lawyer to substantiate this claim, the King offered to take him to meet his ancestors. That was when the defence asked for a medical examination to assess whether the King could be held responsible for his actions.

Mizele was far from being a first-time offender. A similar incident had occurred in 1996, under Mobutu, when employees from the water board went to the Royal Court to chase up some overdue bills. They were beaten and held hostage and in the stand-off that ensued between the court officials and soldiers sent to free the water board workers, one man died. The King was sentenced to fifteen years in prison, but he was out in no time, one of the detainees who scampered from their cells when Makala's guards abandoned their posts during the rebel takeover of 1997.

He had returned smoothly to his old activities and claimed to have been in touch repeatedly with Kabila. The former rebel chief, King Mizele's lieutenants explained, had promised to hand over power six months after toppling Mobutu in exchange for their previous support. The royal court had decided to march on the radio station only when it became clear that Kabila was not planning to honour his commitments. Given Kabila's readiness to make promises to all and sundry in the run-up to his seizure of power, the audience found this part of the story easy to believe. But the jeers and boos began when the xenophobic nature of King Mizele's fantasy state emerged. The Kongo Kingdom was to unite King Mizele's own Bas-Congo province with Kinshasa and the province of Bandundu. Tribes from these three western regions—and in particular, from Bas-Congo—would enjoy special privileges. Those signing up for the Royal Army must come from within the area, said the King, as he did not want 'mercenaries' in his forces. 'Outsiders' were invited to leave Kinshasa of their own free will.

In the decades of Mobutu's rule, the Congolese had seen ethnic cleansing used repeatedly as the Leopard ruthlessly stoked up tribal hatreds for political gain. Scarred by the massacres and expulsions, Kinshasa's population did not want to pass that way again. For those born outside the favoured three provinces, the Kongo Kingdom only promised a repeat of old nastiness. When the King and his personal secretary were condemned to twenty years in jail, with lesser sentences and fines for other notables, there was a general sense of relief. 'We tried to argue that because the King was away in Bas-Congo at the time of the shooting, he could not have given his followers any direct orders or be held responsible,' said one of the unsuccessful defence lawyers. 'But the authorities decided that having told Kabila to quit power in the first place was responsibility enough.'

In an attempt to prevent any further resurgence, the authorities have done their best to wipe out all evidence of this national embarrassment. The Royal Court's activities have been outlawed, the family home King Mizele used as his base seized and converted into a police station. The King's personal secretary died in prison but Mizele himself, placed in the wing reserved for those convicted of military offences, seems to be thriving. He has put on weight, is said to be cheerful and is given to telling visitors he soon expects to be a free man. It is not clear whether he hopes to be pardoned or is planning to take part in one of the all-too-frequent escapes in which, even the chief warden admits, Makala specialises. Perhaps King Mizele's cheerfulness is based on the understanding—the same realisation that made both Mobutu and Kabila belatedly sit up and take notice—that there will always be those exasperated enough to listen to a siren voice speaking of mystical vocations, ancient rights and a brave new world.

King Mizele's thwarted citizens had looked to the past for relief, weaving their dreams around the folklore of an antique kingdom. A music rehearsal in the heart of Kinshasa revealed how many Zaireans turned their gaze inwards to escape reality, resolutely embracing the trivial in their quest for self-fulfilment.

It was hard not to wince as a series of electronic whines shrieked from the stage, followed by the monotonous 'Hallo, hallo, testing, testing, hallo, hallo'. Wenge Musica 4X4, as the pop group called itself, was trying out its system ahead of a concert scheduled a few days hence. The equipment was rudimentary, the amplification turned too high and the result a fuzzy roar in which voices and instruments all blended into one painful, deafening mess.

It was a shame, because there are few sounds sweeter to the ear than the music known across Africa by the generic term 'Lingala'. If Congo has failed in most sectors, music must qualify as its one, most glorious exception. Across the continent and in the Afro-Caribbean nightclubs of Paris, Brussels and London, fans snub home-grown bands to dance to the lilting melodies coming out of Congo's slums.

The mystery is how conditions so depressing can give birth to tunes so infectiously light-hearted, so innocent in tone. But somehow they do. As a music expert once wrote, if the critics' jibe that Congolese guitarists often only play three notes has an element of truth to it, the fact that those three notes have managed to entrance a continent for more than thirty years is something of an achievement. Formulaic though it may be, Lingala is Congo's greatest export, its commercial success the most reliable escape strategy ever made available to its youth. The goal is to be recognised by a promoter scouring the hundreds of tiny nightclubs in Kinshasa, flown to Paris or Brussels to record a first cassette, break into the international music scene and—following the pattern set by such stars as Papa

Wemba, Koffi Olomide, Tabu Ley Rochereau and the late Pepe Kalle—start a new life abroad, only occasionally returning to Kinshasa to perform for grateful fans back home.

Wenge had been one of the latest groups to go through that routine, fulfilling most of the industry clichés in the process. Mirroring the country's political parties, Congolese bands have a tendency to fracture within split seconds of forming, as the most talented members vie for the limelight. Each dissident faction, realising the importance of the recognition factor, then claims the original band name. And so the offshoots confusingly proliferate. What had started out as good old Wenge Musica now came in four rival versions: Wenge Musica 4X4—that day's performers—Wenge Musica Maison Mere, Wenge Musica BCBG, Wenge Musica Kumbela and Wenge Musica Aile Paris. No doubt there would soon be more.

As dusk fell, and a flock of herons flew over the white-washed compound in neat formation, the band was practising its moves, the desire for clarity constantly losing the battle against the quest for added volume. The cost of the tickets for listening to the stop-start renditions of Wenge's hits was minimal, but it was still too high for many of the fans milling outside the open-air venue. The balconies of nearby apartments were crammed with people watching for free and despite organisers' attempts to shoo them away, the phaseurs were out in force on the corrugated roofs of nearby shacks.

Wenge Musica 4X4 seemed to be following the recent trend of downplaying the role of griot—the angelic voice which traditionally sang the king's praises—in favour of the fog-horn voiced 'animateur', who was once limited to shouting out one-word choruses. At his bidding, the performers were now standing in line, legs bent. A pelvic thrust was passed from one singer to another like a bad case of the flu, until six hips were grinding in unison. Then the group suddenly fragmented, each youth wheeling away in apparent confusion, to reassemble in a different formation.

While Wenge's musicians practised, a different kind of choreography was becoming apparent in the young men arriving to help with the rehearsal. Almost willing the crowd to watch them, each crossed the floor alone, sauntering the length of the compound with the diffident self-consciousness of court débutantes, heads high, toes turned outwards, shoulders rolling. There was nothing spontaneous about this gathering. Each wore at least one item of clothing that could qualify as 'tape-à-l'oeil'—designed to leave its image lingering on the retina—a white shirt whose collar wings fell to nipple-level, a pinstripe jacket with a giant diaper pin in its lapel, a black fishnet T-shirt, a drawstring top with one hood in front and one behind, a top and pair of trousers in ice-lolly colours bright enough to make the teeth ache.

Pronounced dead by overly-blasé Congolese radio presenters years ago, 'La Sape'—central Africa's equivalent of the Mod movement—was clearly alive and well, I noted with approval, albeit surviving in straitened circumstances. Proud of their status as fashion victims, a new generation of 'sapeurs' had turned style into a form of near-religion (dubbed 'kitendi'), complete with its 'grand priests'—the classiest of dressers—and its 'deities'—the international designers. The show, evidently, was still going on.

An abbreviation of Society of Ambiencers and Persons of Elegance, La Sape as a movement was actually born across the river in Congo-Brazzaville in the 1970s. But it was in Zaire that it really made its mark, moving hand-in-hand with the explosion of the Lingala music phenomenon onto the international scene and fuelled by the birth of a monied urban elite who had travelled, shopped abroad, and knew their Yamamoto from their Montana, their unstructured jacket from their deconstructed suit.

As bands signed recording contracts in France and Belgium, their members hit the designer shops of Place Vendôme and Place Stephanie, returning to Kinshasa with suitcases full of 'griffes' (designer labels) to show off. Fans of rival bands would compete to see who could look cooler, perfecting dance techniques that allowed them to show off their socks on the disco floor, or display the crucial silk labels on the insides of their jackets. The biggest star of all, Papa Wemba—who enjoyed the jaunty title of 'Le Pape de la Sape'—spearheaded one craze after another with his on-stage appearances.

There was the time of the three-quarter length trousers, the time for braces, the time when Jean-Paul Gaultier was all the rage.

The movement, I knew, had gone into something of a decline with the general drying up of disposable incomes. The death of Niarkos, a famous Kinshasa mobster who rivalled Papa Wemba for narcissism, had dealt it another blow. Yet here they were: the shirts looked a little grey, the jackets far from new and the battered shoes were the biggest give-away, but this flock of down-at-heel young peacocks were keeping up appearances nonetheless.

'Of course it's still alive,' snorted the man known as 'Colonel Jagger'. 'If anything, La Sape has just become part of the mainstream, it's been vulgarised. Government ministers wear couture and have sapeur hairstyles. Just look at the young church pastors—even they now assert themselves.'

Asserting oneself ('affirmer') is one of the key concepts in La Sape's vocabulary, ranking in importance alongside understanding how to 'débarquer'—make an entrance (never, but never, to go unnoticed)—and knowing how to walk. A sapeur's walk is an art form in itself, a mixture of swagger and stroll as individualistic as a graffiti artist's tag. 'Do you remember John Travolta's way of walking in *Saturday Night Fever*? Well, we were doing that long before he did,' said Colonel Jagger. 'You lollop, you almost dance. It's each man's way of standing out from the crowd.'

Recognised as a key proponent of La Sape, Colonel Jagger, manager of the rival band Viva La Musica, nevertheless proved a slight disappointment on first encounter. Dressed in a simple black T-shirt and jeans, this quietly spoken and rather sombre individual had none of the flamboyance I had come to expect. Ah, but that was where I was mistaken, said Colonel Jagger, when I confessed my surprise. 'This may be understated, but it's still La Sape. These are Weston shoes, Ferré jeans and the T-shirt is by Gaultier. All in all, this outfit probably cost over £1,200.'

We were in the heart of noisy, smelly Matonge—'my Matonge' as Colonel Jagger referred to it. But the pink-walled house was situated in an unexpected enclave of peace, hidden down an acacia-lined avenue. In the street, urchins were playing football and neighbours were sitting chatting quietly in the trees' shade. They watched us with interest, but were careful to keep a respectful distance from the Colonel, one of Kinshasa's acknowledged VIPs. And keeping your distance, establishing some personal space, as it turned out, was a principle that went to the very heart of La Sape, along with bitter contempt for slavish imitators ('suivistes') and those with money but no sense of style ('taureaux'). 'Our slogan is "No indiscriminate contact" (Pas de contact avec n'importe qui),' said Colonel Jagger. 'It means we keep our distance from the police, the authorities and we don't get mixed up in politics. We keep away from those people because they don't understand us. They go crazy when they realise that someone with empty pockets is going around in an outfit costing 12,000 French francs (£1,200).' Those whose aspirations had been stifled all their lives had pushed sartorial elegance to a point where it became far more than self-indulgence. It became a mission.

Much of the movement's original inspiration came from the first films shown in Kinshasa. During colonial times, the Belgians would send lorries into what were then the 'indigenous quarters', set up their projectors in the open air and screen movies for the entire neighbourhood. The adventures of the Three Musketeers, with their swash-buckling costumes, and the black-and-white thrillers of the 1940s and 1950s, with their sharp mobster outfits, seemed the epitome of Western cool. Later on came borrowings from the British pop scene. The colonel took on the name of his favourite rock star, Mick Jagger, and acknowledged his admiration for Bryan Ferry, 'my favourite Englishman'.

There were more recent signs that La Sape was being infected by the 'slob' look embraced by America's blacks, all outsized jeans, baggy dungarees and shorts that drop to calf level. But Colonel Jagger, who dismissed the style as 'the white man's look', remained a conservative, with a philosophy bordering on the austere. He stressed the importance of cleanliness, preached against violence, abhorred hard drugs ('if you use hard drugs, you get dirty, so you can't be a sapeur') and shaved his head once a week to avoid a messy hairstyle.

The sapeurs fancy themselves the best dancers in town and are often the players who decide when a particular disco step has outlived its interest and it is time to adopt a new one. The Kwasa-Kwasa, the Kotcho-Kotcho, the Otshule: the dances are born, sweep across the nightclubs of Africa and Europe and then mysteriously disappear, replaced by a new style that involves using the hips more, perhaps, a slightly different rhythm, or moving the foot and arm in tandem.

The crazes are not without a sense of political and social irony. The Etutana dance, based on the principle of rubbing yourself vigorously against your partner and with a chorus of 'ça c'est bon' ('it feels good'), was a reaction to the AIDS awareness campaign which was trying to persuade young Congolese to stop having unprotected sex. More recently, the Ndombolo has been causing a stir. Said to have been invented by the street kids of Kinshasa, it involves spreading the legs far apart, bending the knees and poking one's bottom in the air. Banned as obscene in Cameroon, the Ndombolo is not very graceful. But then, it's not meant to be, because the Ndombolo combines a crude imitation of sex with mocking mimicry of the gait of the overweight Laurent Kabila. This is one dance that did not originate with the sapeurs, who consider it below their dignity. 'A sapeur will never, ever dance the Ndombolo,' said Colonel Jagger. 'Gyrating your hips is fine for women, that's our view. But a sapeur moves as little as possible, just enough to show off his trousers or his shoes. If you're wearing a nice outfit, you obviously don't want to break into a sweat.'

Ask a sapeur about the motivation behind the phenomenon, and he will usually mention a desire to prove to the Europeans, who brought clothing to central Africa, that they could be beaten at their own game, that the once-naked savages had become cooler and more elegant than their dowdy colonisers. But another factor was the desire to react against the stylistic monotony of the Mobutu years, when 'authenticity' led to the outlawing of Western dress, ties were regarded as subversive and the ghastly 'abacost' jacket was supposed to hang in every loyal citizen's wardrobe. For a population known for its love of display, few decrees could have been more demoralising.

'For twenty years people here wore a uniform,' recalled Colonel

Jagger. 'We were the only ones who refused to do so. At concerts sapeurs would be beaten up for wearing suits. It was a way of saying "no" to the system, of showing there's a difference between us and everyone else. A way of feeling good about ourselves.' Once the sapeur had embraced that lifestyle choice, days off were not permitted. 'Sapeurs don't dress for other people. They dress for themselves. And in contrast with most people, who dress up at the weekend or to go out, they dress smartly every day of the week.'

But wasn't all this a rather trivial way of expressing revolt? In other countries, frustrated young men took to the streets or got involved in politics. In Kinshasa, the generation holding out hope for the future was busy fussing about the colour of their socks. Wasn't this a waste of energy better channelled elsewhere? 'It's easy for you to talk. But the older generation here has fenced off the world of politics. This is a world where you can't go out and shout in the street, where you suffocate, because there is no room to breathe. I have no weapons, so instead I create a world of my own,' explained Colonel Jagger.

Certainly, when you considered the practical difficulties involved in being a sapeur today, as opposed to the years when money was still washing around the system, the struggle to 'affirm' oneself acquired a near-heroic quality.

A pair of good shoes started at $100 in Kinshasa, almost equivalent to the average yearly per capita income logged by UN agencies. The clothes displayed in the boutiques of the Hotel Intercontinental, expensive even by European standards, were well beyond most locals' reach. Unable to actually buy the Versace jackets, Paul Smith shoes and Gianfranco Ferré trousers they so longed for, the sapeurs depended on their friends—especially those abroad—for loans and swaps. 'Most of us rely on trading items between friends rather than outright buying,' said Colonel Jagger. 'There's a certain solidarity. I know which of my friends has money at any given time and we help each other out. We tighten our belts. But either you're a sapeur or you're not. It's not a question of money. It's a question of taste.'

Talking to the melancholic Colonel, I was suddenly overwhelmed by that sense of tragic waste, of crippled potential, that so often sweeps over one in Africa. This articulate, subtle man was no longer young. He had reached the age when most men have relegated an obsession with jean brands and fancy waistcoats to the mental drawer where they keep their motorbike manuals and collection of Bo Derek posters. Yet here we were, discussing shoe makes and dancing styles with the seriousness a Buddhist would devote to meditation techniques.

And then I remembered an excerpt from *The Road to Wigan Pier*, in which George Orwell wrote about the spending habits of the poor, the tendency of the bored, miserable and harassed to fritter their wages on chips and ice-cream instead of the dull, wholesome food that would keep them healthy. Middle-class puzzlement missed the point, suggested Orwell, for being able to 'treat yourself' was the only thing that made such existences bearable. La Sape, I realised, was that principle seen through to its philosophical conclusion. Spending your money on a luxury rather than a necessity was part of what kept you human, as essential to a sense of self-worth as the smear of lipstick on the face of a pensioner. Acting the dandy in modern-day Congo was like playing the gourmet in a concentration camp. The harder finding a Comme des Garçons shirt became, the more convincingly its eventual wearer proved he remained master of his fate.

'Papa Wemba, Niarkos and I, we brought the young people here hope—we made them realise that you didn't have to be the son of a rich man to make it,' said Colonel Jagger. 'They regard us as role models. No matter how poor, they aspire to one day dressing like us. Even a boy in the street here will know who his favourite designer is.' I must have been looking sceptical, because Colonel Jagger called one of the urchins playing football in the dust over to prove his point. 'Go on, ask him.' 'Who's your favourite couturier?' I said in French. The boy shuffled his bare feet, picked at his filthy shorts and looked down at the floor. 'I don't know.' 'He doesn't understand the question,' said Colonel Jagger tolerantly. 'Griffe oyo olingaka mingi ezali nini?' he translated, and at the word 'griffe' the boy's eyes lit up in immediate understanding. 'Versace for jackets and Girbaud for jeans,' he answered, without a moment's hesitation.

The sapeurs had managed to keep their dreams alive, skating around the edge of despair without tumbling in. But there were those who had seen their complicated fantasy edifices come crashing down. Standing in the rubble, they gazed around them with clear eyes and shuddered at what they found.

Most of the expatriate community seemed to fall into that category. Often they were the children of Belgian administrators, Greek businessmen or Portuguese shopkeepers who remembered frightened mothers standing guard with machetes during the civil disturbances of the 1960s and listening wide-eyed to tales of lions prowling outside rural compounds. Or they were Europeans of humble origins with grandiose ambitions, who thought they had discovered in Africa the freedom to cast off the shackles of class and prejudice and reinvent themselves.

Congo had been a home that once offered warmth, laughter, endless possibilities and the automatic respect accorded a white man in Africa. Europe seemed alien territory where aspirations were cramped, relationships strained. Initially convinced they possessed the intuitive understanding that would allow them to solve the riddle, beat the Congolese conundrum, they had stayed on after each outbreak of violence. 'This country is like a woman,' a Belgian lawyer lamented. 'She cheats you once and you forgive her and come back. Then she cheats you again and you forgive her once more. She keeps cheating and you keep coming back.'

Now their businesses were barely ticking over. Their numbers had shrunk as, one by one, their friends packed up and left. Resigned to keeping out of politics, they had realised they could not make money either. When they were arrested or threatened by the authorities, they found precious little sympathy at their embassies, whose young diplomats regarded them as unreconstructed colonialists largely deserving the treatment meted out. If they were still universally hailed as 'patron' ('boss'), the title had begun to grate, the nature of their tiny ghetto had become manifest. They now knew themselves doomed to be aliens by virtue of their skin, fixed in the aspic of paternalism.

Unlike most Congolese, they enjoyed the luxury of choosing where to live. But they were trapped in a different way. Uneasy and ill-defined, they were the mirror images of the Congolese exiles trying to start new lives in Paris and Brussels. Souls in limbo, they knew they could not make things work in Congo but had nothing left to give to a European continent whose cold efficiency chilled them. When more recent white arrivals, in boorish expatriate mode, raged against the perfidy of the locals, they concealed their anger. Their tragedy was that they loved the place, but no longer expected to be loved in return. They had come to share with Mobutu the quality of obsolescence. 'We are like dinosaurs, dying off one by one,' acknowledged the lawyer. 'We feel so involved, but we are utterly marginalised, incapable of dictating events.'

On the banks of the River Congo, an hour's drive east of Kinshasa, before the road passes the fishing village of Maluku, I found one of the most poignant members of the breed. Strolling through groves of avocado, grapefruit, lemon and lychee he had planted and nurtured into luxuriant life, patrolling warm brown pools teeming with fish, Daniel Thomas was taking stock of a lifetime of labour destined to leave him empty-handed in what would all too soon be his old age.

His glade exhaled lazy peace, sun-drenched contentment. Troupes of guinea-fowl, their bobbing heads the bright turquoise of a tropical beetle, picked their way across the green lawns, butterflies wafted over the yellow hibiscus and a widow bird looped from bush to bush, its languid tail dipping. It was hot, and while the staff prepared an open-air barbecue, Thomas's dog determinedly dug a hole in the lawn, where it sat panting, cooling its belly on the exposed earth. From the river came the sound of a barge pushing logs cut many, many miles upstream, somewhere in the equatorial jungle where Mr Kurtz gradually lost his reason. But its chugging progress was dwarfed by an expanse as vast as his dreams: the cloud-dotted sky meeting a mother-of-pearl sheet of water in a horizon that was no more than a shimmer of grey-blue.

From Thomas's farm the sometimes imperceptible curve of Malebo pool made itself manifest. To the right, across the water, lay a thickly forested island. To the left, you could see the slim tower of Nsele, where Chinese workers built Mobutu an ornate pavilion and the single party system was born. Behind it, the glints from the skyscrapers of two turbulent African capitals. The heat haze was pierced by the odd plume of smoke from a farmer's fire, thin threads of white trailing across the rolling, pea-green landscape that must have looked so terrifyingly alien to the eyes of Stanley and Brazzaville, accustomed to the cosy fields and neat hedgerows of Europe.

'There was nothing here when we came in 1976, just brush to be cut down,' said Thomas. 'Now look at these mango trees.' He bent to point out the little buds in the thick green foliage, shaking his head with wonder at the fertility of the soil. 'They're flowering again, and we've only just finished eating the last crop, which were mouth-watering. What a feeling, to pick fruit from something you have planted yourself! There's nothing finer in the world.' His face was the colour of baked terracotta and it had the glazed quality of someone who had spent his entire life working outdoors. His teeth, stained by cheap local cigarettes, pointed in a variety of unconventional directions. But his blue eyes, though tired and watery now, still had the trusting innocence of a child. Which seemed appropriate, for Thomas himself admitted he had been truly infantile in his slowness to learn the painful lessons of experience. Bewitched by his vision of an African Garden of Eden, he had been like a toddler who tumbles, gets up, is knocked down a second time, falls once again, staggers back on his feet, only for the whole process to be repeated once more. It was hard to know whether to admire his commitment or dismiss him as a fool. Either way, you had to marvel at his energy.

Thomas and his wife, a pale, melancholy woman, had been looted not once, not twice, but three times in eight years, an escalating ladder of theft and destruction that had worn away the huge store of optimism they had brought to the country. In fact, if you counted Zaireanisation, you could argue that the couple have been ripped off a total of four times in the land they once wanted to make a home but now talked of leaving.

He had come to Zaire in 1970 as a construction expert speaking with the twang of rural northern France, brought in to set up and run factories producing high-quality tea in the eastern Kivu province for export to the Common Market. Now associated with sprawling refugee camps and rebel uprisings, Kivu was then a peaceful province of green hills and misty mountain ranges. The madness began, Thomas was taken over by the farmer's passion for the soil. 'The climate, the people, the land. It was paradise on earth: I bought 150 hectares outside Bukavu and lived like a king.'

Then Mobutu introduced Zaireanisation, and foreign-owned farms, factories and businesses were allocated to cronies with little desire to get their hands dirty. 'Zaire', as Thomas put it, 'began to self-destruct'. The tea project took only three years to collapse. Expecting to lose his own land, Thomas donated it for free to a Franciscan order and moved to Kinshasa to start again. He bought the 114-hectare site near Maluku and raised Barbary ducks, which proved profitable until most were killed by bad feed from the only suppliers. 'I sent the feed for analysis and was told it was full of sawdust and coffee grounds. The ducks died of hunger with their stomachs full.' So Thomas taught himself how to graft fruit trees and built up a herd of cattle and sheep with which he supplied Kinshasa's Moslem community for the yearly El-Khadir festival.

Digging a three-kilometre channel to divert water from a nearby stream, he created five pools on the river bank and stocked them with tilapia and capitaine, the most popular species of fish in this part of Africa. Anglers who could tolerate the sauna conditions at the river's edge would come and catch their own. The couple built paillottes—thatched awnings—and word spread of a new place to lunch over the weekend. On a Sunday afternoon, they sometimes found themselves running to serve roast fish, barbecued lamb and home-made ratatouille to 300 guests.

The concern was thriving by 1991, when the first round of army-led looting swept across the country. Thomas estimates that he spent 100,000 French francs (£10,500) repairing damage done to the property and replacing stock that time. Two years later, when the riots and pillaging broke out again in Kinshasa, the farm was more seriously affected. 'We had all the kitchen and farming equipment stolen and lost 280 sheep. The damages totalled about 700,000 francs (£74,000).'

But the incidents were dwarfed by what happened in 1997 in the run-up to Kabila's seizure of power, just as the couple were about to harvest their fish. Marooned in Kinshasa, Thomas had fretted about the livestock on the farm, situated worryingly close to the road Mobutu's soldiers were meant to defend against the oncoming rebels. Finally, he set off with a stock of cash and methodically paid his way through twenty military roadblocks, until, at the last checkpoint, he was held hostage while the gardes civiles debated taking his car.

On his release, he found the scene he had dreaded. The paillotes had been burnt down, the farm stripped bare, the herds of sheep and cattle shot and for the first time the sluice gates had been opened and the pools emptied of all their tilapia and capitaine, dumped on a local market. The losses this time were a crushing 1.2 million francs (£126,000).

If businesses looted in Kinshasa through the years were hard put to identify their attackers, for the Thomas couple there was no such comforting anonymity on offer. Depressingly, the people who led the soldiers to the farm each time were local villagers. Far from regarding the farm as a project worth encouraging, or at least tolerating, for the investment and employment it might bring to the area, they monitored the farm through the years like schoolboys watching a ripening fruit, waiting for the moment when a breakdown of law and order would provide the cover for some neighbourly appropriation. 'It's always the same hard core that incites the other villagers and brings the soldiers here. We know who they are, we even know their names. But they've never been punished and they never will be,' said Thomas.

He had not quite managed to kick over the traces. He had restocked the fish pools and rebuilt the paillottes, though customers willing to risk the journey had become a rarity. In his stained shirt and old trousers, he toured the banks with cutters in hand, exclaiming like some modern-day Andrew Marvell over the delights of his jewel-green kingdom. He could not begrudge the money he had lavished on the farm over the years for, like the lawyer, his feelings for the land were those of the romantic lover. 'It's like a beautiful woman, you don't count what you've given it.'

He appeared to harbour little rancour, attributing the repeated pillaging to the hunter-gatherer instincts on which the Congolese relied for survival until so very recently. But something in this man who attributed his career to 'eternal optimism' appeared to have snapped this time, perhaps overwhelmed by the realisation that those around him never regarded him as anything more than just another white colonialist to be taken for a ride at worst, deferred to at best. Since the last looting, he said, the couple have given up their long-held hopes of building something permanent and spending their last years in Congo. 'The punch just isn't there any more. It's obvious that nobody here understood what we were trying to do. We have no more hope. All we want is to be left in peace.' One sensed gentle pressure from his wife, who had not been well. Her face did not have his childlike gleam, it looked bleached by disappointment and betrayal. 'We put our hearts into this place twice and our hearts were broken,' she said. 'We have been broken. You try and you try and you try, and then you just run out of energy.'

Now in their sixties, the two found themselves caught on the horns of a financial dilemma common amongst the expatriates. Having always assumed that they would spend their retirement in Congo, all their savings were invested in the farm, which would be impossible to sell in the current political climate. They no longer believed they could build a future here, but with the farm all they had to show for their efforts, they risked the miserly existence of the pensionless in Europe if they left.

Vaguely, Thomas talked about finding a manager who could be trusted to run the farm in his absence and send the proceeds on to Europe. But even the 'eternal optimist' struggled to believe in the existence of such a man. The alternative scenario was all too easy to imagine: falling standards in the restaurant, livestock that mysteriously disappeared and then one day the missing manager, the faked accounts, and a clearing returning slowly to the bush.

And behind the financial problem you sensed the deeper, more philosophical quandary. The wilderness that had intoxicated Conrad's anti-hero had transformed the lowly construction worker into something rich and strange. Even if in the end Thomas's dream had shrivelled, for a time he had ruled over his verdant empire, revelling in the pleasure of moulding the landscape, watching his seedlings stretch towards the sky and his fish grow fat in the warm brown waters. After nearly half a century in Congo, Thomas, like his fruit trees, had become a strange type of hybrid, neither European nor African. It was impossible to imagine the battered lord of this riverbank kingdom hobnobbing with the locals in a French village.

The prospect, he reluctantly acknowledged, horrified him. 'I feel ill at ease in Europe. I find people have let themselves go. Maybe they've had it too easy but they seem to have lost their initiative. I know returning won't be easy. But I'm faced with a Corneillean choice—a retirement here, full of problems and hassle, and one over there, in alien surroundings. It's my own fault. I should never have started, let alone stayed so long.'

Driving back to Kinshasa in the golden light of early evening, the mood in the car was meditative. 'I'm willing to bet that in a year they'll still be there, saying they're about to go but not quite picking up the courage to leave,' said the lawyer. 'In their heart of hearts, they know they'll never find the manager they're looking for. I've seen it so many times. They can't bear to leave and they can't bear to stay.'

On the highway back into town, groups of heavily armed soldiers were materialising out of the long grass with alarming suddenness. The president of Zimbabwe was expected in town, so security was being tightened. At the roadblocks, soldiers were shouting in angry Lingala at the creeping cars, pocketing bribes and ordering passengers out to be searched. The noise and nastiness shattered the lazy Sunday mood. We were being reminded who was boss.

Queuing to have my bag inspected, I looked up and spotted the Thomas couple, returning to their home in Kinshasa after a day at the farm. They were squeezed into the front of a small green pick-up, with a Congolese girl and a branch of bananas in the back. After Zaireanisation and three rounds of looting, the military roadblocks represented a more subtle form of economic looting. They would ensure the number of visitors to their farm and restaurant remained at a dribble.

They had either not understood the soldiers' orders to descend or could not be bothered to obey, and were both still behind the wheel. The setting sun was in their eyes and Thomas was squinting into the light, his face creased into well-established wrinkles. Beside him, his wife looked fatigued. They spotted me at the roadside and smiled a moment before being waved through, escaping the frisking. In this most trivial of encounters, at least, luck had been on their side.

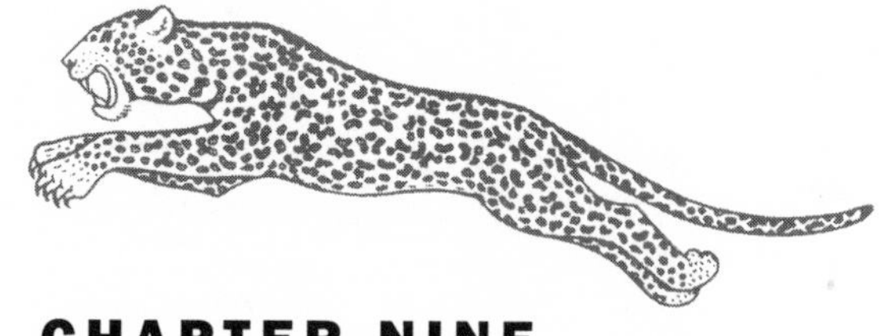

CHAPTER NINE

I get by with a little help from my friends

'I cannot live outside the budget. Where would the money come from? It's just not possible. My enemies say anything to bring me down.'

—Mobutu Sese Seko

There is something rather touching about the figure of the late Erwin Blumenthal, the German banker who briefly tried to put Zaire's finances in order and found the experience so terrifying, the rules of the game so far beyond anything he had hitherto encountered, he ended up sleeping with a pistol under a pillow, braced for imminent assassination.

In the Washington institutions he passed through, he does not seem to have left many friends behind. Describing him, 'cantankerous', 'punctilious' and 'prudent' were the adjectives that sprang to the lips of former colleagues. 'He was a petty bureaucrat, and petty is one thing Zaireans have never been,' sniffed one. 'He was used to the clockwork efficiency of the Bundesbank and just couldn't adapt to Africa,' was another's assessment. Most damning of all: 'He was very, VERY German.'

No, Blumenthal was definitely not the stuff from which heroes are made. But that very Prussian meticulousness, wrapped in a thick layer of obstinacy, was what allowed him to leave his small mark on history. This bureaucrat who so got on everyone's nerves managed to pull off a double *coup*. Not only did he expose to public view the financial machinations of Mobutu and his 'Big Vegetables'. Far more significantly, he made it impossible for the West to ever pretend it was not aware of what was going on, bringing it face to face with its own hypocrisy.

As the years passed and Zaire was progressively pauperised, the bystander might be forgiven for concluding that the outside world was being kept in blissful ignorance of Mobutu's venality. How else to explain the level of aid the country continued to enjoy? Between the start of the Zairean economic crisis in 1975 and Mobutu's departure in 1997, Zaire received a total of $9.3 billion in foreign aid. Between 1975 and 1984 the sums averaged $331 million a year, rising to an annual $542 million from 1985 to 1994.

Leading the way were the International Monetary Fund and the World Bank, the two institutions whose charters were first drawn up in 1944 with the backing of President Franklin Roosevelt, who was convinced that financial crashes, economic depressions and currency disorders lay at the heart of two world wars and that free trade, reconstruction and the development of backward economies were the keys to international peace.

Since their beginnings at a hotel at Bretton Woods, New Hampshire, these two institutions have come to play a crucial—and controversial—role as 'catalyst' lenders in the Third World. The amount of money they themselves lend may be quite small, but as a symbolic vote of confidence, a World Bank or IMF agreement is the signal governments and commercial banks look for before going in.

Given that the World Bank and IMF remained on working terms with Mobutu from soon after his takeover until the early 1990s, despite the repeated failure of the economic stabilisation programmes launched in Zaire, the natural assumption might be that they were unaware of the ghastly truth throughout these years. In fact Mobutu's foreign financial backers knew all too well what was going on. If the surreal anecdotes of expatriate businessmen, the press accounts of the president chartering Concorde to take his family to Disneyland were not enough, their own staff had been warning them for decades of what was happening, as the report drawn up by Blumenthal made clear.

Despite his colleagues' dismissive comments, Blumenthal was by no means a newcomer to Africa when he was sent by the IMF to take up the post of director of Zaire's central bank in 1978. He had worked in Tanzania and acted as adviser to Moise Tshombe, the

Katangan secession leader who briefly served as prime minister. But it was fourteen years running the Foreign Affairs section at Germany's central bank that marked him out as a likely steady hand on Zaire's erratic tiller.

Nominally, the dispatching by the IMF and World Bank of foreign advisers to work inside key institutions is justified on the grounds that weak, post-colonial administrations lack technical expertise. In Zaire's case, the placing of experts in the central bank, customs service and Finance Ministry betrayed concern over the leakages that risked sabotaging all attempts to revive the economy.

Blumenthal lasted only until 1979 before giving up. He did not immediately rush into print—as director of the central bank, he was bound by professional confidentiality. But three years after his abrupt departure an institution attached to a creditor government asked him to assess the likelihood of Zaire repaying its foreign debts—by then a crippling $5 billion. The report he produced is an idiosyncratic piece of work. Written for internal consumption and mercifully devoid of the diplomatic circumlocutions normally used in World Bank and IMF publications to skirt such unmentionables as 'graft' and 'sleaze', it attempts neutrality, only to lapse into fury. It has the freshness that comes of the release of pent-up exasperation.

Despite massive external assistance, Blumenthal begins, all Zaire's indicators show a disastrous five-year slide. Given this fact, he asks, why do the IMF and foreign donors insist on renewing their loan agreements? In theory, he acknowledges, with its extraordinary reserves and a president ready to promise to meet IMF conditions, Zaire is an investor's dream. But that is to ignore the only thing that really matters, he warns: 'There has been—and there still is—one single major obstacle wiping out all prospects: the corruption of the team in power.'

He provides a random list of incidents that occurred during a posting which, ironically, coincided with what Mobutu declared 'The Year of Raising Morality': furious showdowns with generals, gun-waving soldiers and government officials demanding tens of thousands of dollars (in cash, of course); the realisation the Bank of Zaire's accounts abroad had first been plundered and then falsified; and, finally, the discovery of 'special accounts' opened by the central bank in Brussels, Frankfurt, Geneva and London in the president's name which, surprise, surprise, never featured in official records.

The report is not without a certain accidental humour. Blumenthal, the professional whose word is his bond, was flummoxed by Mobutu's capacity to tell the most outrageous lies without pausing for breath. You can sense the hilarity that must have exploded in the presidential palace after each of his worthy, pompous remonstrations over yet another monstrous presidential fib.

With the president's blessing, Blumenthal puts Mobutu's uncle, who has run up massive debts, on a banking blacklist. A few weeks later the central bank pays the uncle $50,000 in cash, blacklist or no blacklist. Mobutu agrees that a massive salary being paid a university professor in Belgium should be slashed, only to quietly restore payments to the man, who happens to be his son's guardian. Best of all are Mobutu's sympathetic noises, his regret, his reprimanding of the central bank governor, when Blumenthal leaves, overwhelmed by the president's quiet sabotaging of his work. 'What an actor!' marvels Blumenthal in his report.

And that is his main point. The corruption in Zaire, he argues, is not a generalised blight, a plague without face or source. Mobutu, he argues, is too feared, too powerful for his minions to be taking these actions without his approval. Yet—you can almost feel his blood pressure rising—the IMF and World Bank were still giving Mobutu's reform plans serious consideration at the start of the eighties, despite the fact that he had kept them scrupulously informed of what was happening. 'None of the Fund and World Bank Officials can be unaware that any attempt at strict budgetary control comes up against a major obstacle: the presidency!' he fumes. 'Any monitoring of the presidency's financial transactions proves impossible. In that office, no difference is made between state expenses and personal requirements. How can the international institutions and Western governments still be putting blind trust in President Mobutu?'

He reaches a final, depressed conclusion: 'There will certainly be new promises from Mobutu and the members of his government and the ever-growing foreign debt will be rescheduled. But there is no—I repeat no—chance on the horizon that Zaire's many creditors will recover their funds.'

Blumenthal's devastating finale was an appendix in which he listed questions and answers put to Nguz Karl i Bond, the former prime minister who had gone into exile. Nguz was later to throw his hand in with the president once again, in one of those political U-turns Mobutu so excelled at masterminding, but at this stage of his life he was in denunciatory mode. He helpfully details embezzled funds, confirms the existence of a bevy of presidential accounts in foreign banks, explains the method Mobutu used to take a cut on sales of cobalt and copper sales, and lists known presidential properties in France, Belgium, Switzerland, Italy and Africa.

But the last pages were the most embarrassing for the foreign allies, because they give a hint of how thoroughly Mobutu, who probably spent more than any other African leader on press relations firms, well-placed political 'friends' and lobbyists in the key Western capitals, had used his wealth to buy support abroad. Nguz names top Belgian civil servants, politicians and journalists on Mobutu's payroll. Former French president Valéry Giscard d'Estaing was among those to benefit from Mobutu's largesse, he claims, which came in the form of diamonds for his wife and priority debt repayment for the French companies in which Giscard d'Estaing's family had an interest.

The report was too juicy not to be leaked: whether this was done with or without Blumenthal's approval is not clear. In the mood of collective amnesia that has developed around the whole episode, IMF and World Bank officials are now wont to play down its significance. 'Oh, there was nothing in it we didn't already know,' one told me. 'In any case, what he mentioned was a fraction of what was really going on,' said another.

But the reason the report was significant was not so much because of the information it contained, but because it ended the cosy arrangement in which the Zaireans knew that the international financiers knew, and the financiers knew that the Zaireans knew that they knew, but everyone could carry on playing the game of credits, conditions, targets and stand-by arrangements with apparent innocence. 'It was a bombshell,' acknowledged one World Bank official. 'The report came out just before we were due to meet a Zairean delegation and I wanted to crawl under the table. I couldn't look them in the eye. What could you say to them after that?'

Bizarrely, Mobutu himself appeared to have played a part in ensuring the report was widely circulated, ordering an emissary to hand a copy over to the Belgian Foreign Minister. One African newsletter speculated at the time that the president was using the report as a form of moral blackmail: too many top Western officials had been complicit in his system, he was effectively telling the international community, for them to stop lending now. Whatever his calculations, his arrogant assumption that funding would continue proved correct, despite the odd hiccup. By the time of the report there had already been four failed IMF stabilisation plans. But the rescheduling of Zaire's debt went on—nine times between 1976 and 1989. It was only in 1990 that the World Bank finally cut off programme funding to Zaire, to be followed soon after by the IMF and bilateral donors.

After Blumenthal's report, there was to be a wearying succession of threats, pleas and pressure on the one side and broken promises, procrastination and bad faith on the other. It was to take eight long years before the two institutions finally reached the same conclusion as the testy German banker had spelled out in 1982: money was not the answer to Zaire's ills, rather, it lay at their very root.

You know when you are entering Bretton Woods land in Washington from the badges. As you approach the cross-section of Pennsylvania Avenue and 18th Street, the men in elegant suits begin to look more cosmopolitan. Skin colours change and you start picking up snatches of conversation in French, Russian, Swahili. But the real give-away are the passes, usually dangling from a metal chain—their ticket into the inner sanctums of the two institutions Roosevelt believed could change the world.

Sandstone brown and tucked discreetly on a side street, the IMF is the smaller of the two, its innards an ugly maze of yellow corridors and small offices. Its staff used to work inside the World Bank building, but were squeezed out as that institution mushroomed, spilling beyond the vast grey-white block to take over a bevy of adjoining tower blocks.

Peeking into the light-filled atrium of the main building, where the fountains drowned out the multilingual babble, I was glad the World Bank, at least, had a certain architectural grandeur. Having worked in so many African countries where its dictums were a matter of life and death to residents, I did not want to be palmed off with some anonymous office block. For the sake of those whose lives had been forever altered, a little magnificence seemed in order.

Recreating how the two organisations acted in Zaire, trying to establish exactly why they chose to ignore their own emissary, should have been easy, but it was not. Under James Wolfensohn, the Australian private banker who assumed the World Bank presidency in 1995 pledging to increase transparency and accountability, country assessments once jealously guarded by World Bank representatives are supposed to be open to the public. But it is unclear whether the new openness applies retroactively. In any case, headquarters in Washington only hold documentation for recent years. Studies dating back a decade or so would have to be unearthed from archives in Pennsylvania and transferred, I was told. That would require authorisation. It was never granted.

So, talking on the apartment balconies or in the neat gardens with the officials who had fallen in love with Zaire, jousted with Mobutu, and finally left with a despairing shrug of the shoulders, it was a question of trying to pull together a spray of scattered personal recollections dating back ten, fifteen, twenty years: an anecdote here, an incident there. The dates did not always match up—memories are selective things—but there were enough consistencies to form a rough picture of the post-Blumenthal era.

Any expectations I might have been harbouring of self-flagellation were to be disappointed. Liberated by retirement or working as consultants for foreign governments, these men were certainly open. But few of these well-educated, highly intelligent individuals—players still helping to make key decisions on which struggling governments received IMF and World Bank aid—felt they had apologies to make forty-five million Congolese, now landed with an unpayable $14.5 billion in debt. If anything, they felt they had done rather well in standing up to Mobutu as strenuously as they did. 'You can't look at it with the values of 1999. You must remember the context in which we were operating at the time,' was a constant refrain. That context was a network of interests that all combined to play straight into the hands of a master manipulator, the student of Machiavelli who had taken the motto 'divide and rule' to heart.

For the Americans, the Afro-optimism of the 1960s had ebbed away, to be replaced by a pragmatic appreciation that it needed Zaire as an ally in the fight to stem Communism's spread. Even before Mobutu became head of state, the White House had signalled its regard by inviting the army chief to meet President John Kennedy in Washington. The White House welcome was to be repeated under every American president through the 1970s and 1980s. The US, using Zaire's bases as the conduit for arms destined for Angola's rebels, was determined to keep Mobutu on board.

For the French, the motivation was different, but no less compelling. Despite its Belgian roots, Zaire had come to be regarded as forming part of what Paris had labelled its 'chasse gardée'—that 'private hunting ground' of African allies whose existence allowed France to punch above its weight in the international arena.

Part of the panoply of French-speaking African countries forming a bulwark against encroaching Anglophone influence, Zaire was a country where French businessmen, shored up by their belief that they—in contrast with the clumsy Anglo-Saxons—understood the

African psyche, hoped to do business. The schools and media propagated the French language, culture and values. In return, Paris assured Mobutu, as it assured all its African dinosaur friends, of its undying support.

For the Belgians, it was a question of maintaining a toehold of influence in a former colony that was still home to several thousand Belgian expatriates. Their safety was of less importance than national prestige: however difficult the scars left by colonisation made the task, Brussels was determined to maintain a historical link that allowed a small, none-too-impressive European nation to count as a significant world player.

All three nations wanted guaranteed access to Zaire's mineral reserves—especially, in the case of the US, the cobalt it needed to produce its fighter jets. And all three were counting, as Mobutu launched into his period of white elephant projects, on picking up some fat commercial contracts. Yet, underlying everything was the strategic question. 'Après moi le déluge,' Mobutu had told Western nations, echoing Madame de Pompadour and, with memories of the terrible 1960s at the back of their minds, they were ready to believe him. 'The notion that only Mobutu could hold the country together grew and grew like a wheatfield,' acknowledged a US official. 'And Mobutu encouraged it.' When Chester Crocker, the former US assistant secretary of state for Africa, raised Mobutu's latest distasteful antics with Alexander Haig and George Shultz, two of the secretaries of state who served under Ronald Reagan, the usual question would surface. 'They would say "We know he's evil, but who else is there?" '

This troika of Western countries was to demonstrate its commitment to Mobutu over the years in solid military terms. When rebels invaded Shaba in the 1970s from Angola, the US organised a military airlift and France parachuted legionnaires into the southern town of Kolwezi. When, in the 1990s, the army rioted in Zaire's cities, French and Belgian troops flew in to police the streets. Nominally, they were there to evacuate their citizens, but in the process they saved the Mobutu regime.

Hand in hand with the military help went the troika's votes within the World Bank and IMF. And Mobutu knew just how to ensure they kept the pressure up, exploiting the fear each of the three harboured that, should they ever dare to express their disapproval too sternly, Mobutu would be pushed into the arms not so much of the Soviet Union, but their Western rivals. 'He played us, and his environment, like a Stradivarius,' recalled a rueful Crocker. 'He would play us off against the French, the French against the Belgians, the CIA against the State Department. If we dared to mention IMF and World Bank concerns it would be: "Do you really expect me to think you're asking these questions of Israel and Egypt? Perhaps I should convert to Judaism." I always looked at a trip to Zaire as hard work, because I knew I'd have to put up with a lot of crap.'

If one ally proved reluctant, Mobutu could always turn to its rival. When aid showed signs of drying up, he would persuade the US to buy six months' worth of cobalt to add to its strategic reserves. But that was not usually necessary, thanks to the quiet influence the troika's representatives enjoyed within the international institutions. Pressure was usually discreetly applied, but on occasions it could be made explicit. When the World Bank was about to cut off relations with Mobutu the then US ambassador to Kinshasa stormed into the Washington office of Kim Jaycox, the organisation's regional vice-president for Africa. Despite his loathing of the Mobutu regime, the ambassador nonetheless felt obliged to warn Jaycox that Washington opposed what he was doing and there would be 'consequences' for the institution if he froze out Zaire. The cut-off went ahead notwithstanding, but the threat helped explain why it was so long in coming.

Selfish 'troika' interests were not the only reason aid continued. Until 1979, when Robert McNamara, the World Bank's hyperactive president of the day, spelt out the concept of 'structural adjustment', in which loans would be made conditional on steps being taken to transform a country's economy, both institutions felt they had no mandate to dictate economic policy to the countries they loaned to. Now regarded by the World Bank as the key factor in granting aid, 'governance issues' were initially viewed as off-limits by officials wary of accusations of neo-imperialism. Even after 1979, macro-economic criteria such as whether a country had a liberalised exchange rate, was keeping its interest rates within the required spectrum or privatising its state sector would take precedence over the question of the head of state's Swiss bank accounts when it came to granting aid. 'At the time you couldn't talk about governance issues or corruption. Our legal department would crack down on you if you tried. It simply wasn't done,' maintained a World Bank economist. On paper at least, Zaire was for many years one of the few African countries that met those macro-economic criteria. The fact that Mobutu's interference made such notional achievements irrelevant was brushed aside. Bretton Woods officials would pay the price for such breathtaking naivety.

But there is another, more insidious reason why the sick relationship with Zaire lasted as long as it did. Ever since McNamara launched his drive to boost lending in 1970, heedless of the debt problem the frenzied credit boom would bring once interest rates rose, Bretton Woods officials have been assessed on their ability to 'push money out the door', to use the World Bank's own phraseology. Projects launched, programmes got under way are, after all, positive, quantifiable achievements in an individual's career, suspensions and cut-offs negative sign-posts on a CV, their effects incalculable and unclear.

In a small local bank, a bad debt comes back to haunt the man who granted it, as someone must eventually cover that loss. With the Bretton Woods institutions, where the lenders are Western governments, bail-out is always assured, a bad debt a sign of good intentions rather than poor assessment. And no one wants to be the official remembered as having 'lost' Zaire, Kenya, Zambia or Tanzania. 'You can never underestimate the inertia of a big institution,' said a diplomat who served in Kinshasa. 'Banks are all about cash flow. They exist to lend money. The World Bank and IMF weren't too bothered where the aid was going or whether it would be repaid. Just as long as it kept flowing.'

Postings are fairly short, which contributes to a vacuum where institutional memory should be. One resident representative comes in from Washington, full of enthusiasm, determined to boost lending to what is obviously a desperately poor government. Three years later he is wiser and far more cynical. He has learned to appreciate how deeply 'governance issues' sabotage every project and finds himself, poverty or no poverty, advising Washington to toughen lending conditions. At about that stage, the resident representative is replaced by some bouncy newcomer who cannot understand why the lending portfolio is so slim and suspects his predecessor of losing focus. The whole cycle starts again, to the vast amusement of the host president, who has seen it all before and knows just when to strike.

And so, thanks to this combination of factors, Mobutu got away with the most outrageous behaviour. One thing almost every official agreed upon was how testing their dealings with the Zaireans could be, how relieved they ultimately felt to move on to new dossiers. 'The Zaireans were the most arrogant people I ever had to deal with', said a former US official who flew regularly to Kinshasa to meet Mobutu. 'They were arrogant and condescending unless they wanted something, and then they were obsequious.' Even Kim Jaycox, not a man who gives the impression of being a soft touch, found the meetings with Mobutu a challenge. 'He was a very formidable, a very wily adversary and he tended to personalise the confrontation. You really did feel you were *mano a mano* with this guy.'

With so much at stake, the famous Mobutu charm could be strikingly absent. He would cajole, bully, threaten and browbeat. His loyalty could never be taken for granted. He liked to show who was boss and was not above demonstrating his independence from his faithful backers with flamboyant gestures of petulance. In 1975, for example, he accused the CIA of plotting his overthrow and expelled US officials. The message was clear: he would not be taken for granted.

Keeping him onside did not come cheap. Roger Morris, responsible for African affairs at the National Security Council under both presidents Lyndon Johnson and Richard Nixon, once estimated that Mobutu received close to $150 million from the CIA during the first decade or so of his regime. Not all that money would have been originally intended for him. John Stockwell, a CIA man who ran one covert operation to destabilise Angola's Marxist government through Zaire, logged how Mobutu creamed off part of any consignment destined for Angola, on one occasion in 1976 casually pocketing the $1.4 million given him by the US to pay off the rebels. Ten years later, a state department official was still being confronted with the same problem: 'We'd mostly stick with equipment as if we sent money we knew it would go missing. But even when we were shipping equipment and gasoline, the Zaireans would steal part of it. I don't think they knew how to do business normally.'

Presidential tactics would range from the impish to the thuggish brutality of a Mafia boss. In Washington, IMF and World Bank officials were always aware that their representatives on the ground were putting more than their professional integrity at stake in taking up a Kinshasa posting. Blumenthal's fear of assassination does not seem absurd when you consider what happened to one Bretton Woods representative who must have displeased the president in the late 1970s. An army unit descended on his home in the normally tranquil diplomatic district and during the prolonged assault that followed, he was beaten up, his wife and daughters raped. Neighbours hearing the fracas called the police, who refused to intervene in what had all the hallmarks of a politically sanctioned raid. 'There was no doubt in our minds, given the nature of the attack and the police's refusal to get involved, that it had been condoned by Mobutu,' recalled a former superior. 'We shipped the family out, and demanded an apology and reparations from Mobutu if the programme was to continue. And we got them, but the damage had been done, of course.'

Willi Wapenhans, the World Bank's regional vice-president for Eastern and Southern Africa from 1976 to 1983, experienced first-hand the strong-arm tactics the president was ready to use. He had gone with his country director to Kinshasa in 1979 to confront the president over $100 million that had suddenly gone missing from Gécamines' foreign exchange earnings, a shortfall that risked bankrupting the copper giant. The two, who had been told by the central bank governor that the president was personally responsible for the withdrawal, were taken to their lodgings, one of the guesthouses in the residential 'village' constructed when Kinshasa hosted an Organisation of African Unity summit. 'It proved to be rather a hazardous exercise. We were expecting to meet Mobutu. Instead we suddenly had soldiers in battle fatigues moving in and surrounding our house and we were effectively being held hostage. The situation lasted forty-eight hours. The central bank governor and Finance Minister would come and visit us from time to time and finally we drew up a memorandum detailing events and told them that if we were not allowed to meet Mobutu we would order our local representative to release it to the key embassies.'

The threat did the trick and a meeting was arranged at which Mobutu explained that the soldiers had, of course, been posted around their house for their own safety, and promised to restore the missing $100 million. The money was, in fact, returned, but soon afterwards the Belgian government passed on to the World Bank president a complaint from Mobutu that the bank was showing an 'inappropriate' level of financial prudence, a view, it was clear, Brussels shared. 'Of course, that wasn't very motivating,' admitted Wapenhans.

Another time, Mobutu engineered the reassignment of a World Bank resident representative by complaining about the supposed 'racist slurs' he had made during a talk with the president. 'Everyone who knew the man in question knew it couldn't be true. I investigated the matter and there was no evidence. But he was removed anyway, as it was clear that all dialogue with Mobutu had become impossible. Those were the kind of tricks he would get up to,' said Wapenhans.

Throughout these years, Mobutu would give the impression of movement, reform and change with a series of government reshuffles, ministerial sackings and central bank appointments. The outside world would assume that lessons had been learned, that reform was underway. But the figure at the centre of the spider's web had not changed, of course.

Nebulous statistics were another great weapon in the Mobutu armoury. Zaire's statistics-gathering apparatus was so inadequate that many figures were based on extrapolations of surveys carried out in 1959, when the country was still under colonial rule. Curiously, the partial export records, the multiple counting of government employees, the fact that nobody, not even Mobutu, knew the actual size of either the population or the army, never stopped the World Bank and IMF issuing hefty reports full of sweeping analysis and confident projections. 'We never really had solid data, because they weren't willing to provide it,' admitted a senior World Bank economist. 'We could never get a good grasp of what was happening.'

It all made cooking the books that much easier as Mobutu focused on his main task, appropriating his 'dotation présidentielle' (presidential endowment). This allowance, which army trucks would regularly be sent to the central bank to collect, irrespective of what the Treasury had officially allocated him, was meant to cover Mobutu's personal security, the costs of his entourage and travel expenses. Somehow, it regularly accounted for between 15–20 per cent of the government's operating budget.

Cleophas Kamitatu, who served as both Agriculture and Finance Minister, acted as unhappy mediator when, in the early 1980s, the Bretton Woods institutions launched one of their periodic attempts to rein in the allowance, which was showing signs of ballooning out of control as work on Gbadolite palace escalated. 'We decided together that $2 million a month should be enough. When I went to see Mobutu and told him, he said: "You're pulling my leg. It's out of the question. I need $10 million. I told him the World Bank and IMF would never agree to that and after a lot of discussion we agreed on $3 million a month, which, after all, added up to $36 million a year.'

Yet, within a week of the Zairean delegation returning to Kinshasa, Mobutu asked the central bank governor for $10 million, citing 'the country's interests' as justification. A month later, there was a request for another $10 million. 'Four months after the IMF and World Bank meeting, he'd already had $36 million, the agreed budget for the year,' marvelled Kamitatu.

It was shortly after this that the first structural adjustment programme went into action, at the end of 1982. A new generation of policy-makers had emerged—that institutional inertia at work again—and there was also a feeling in Washington that with national bankruptcy now a concrete threat and his regime unchallenged on the political front, Mobutu might see the need to knuckle down. For three years the calculation seemed the right one as, under the tutelage of Prime Minister Kengo Wa Dondo, Zaire set in place a reform programme regarded as a model of its kind. The currency was devalued, marketing monopolies were broken up, public sector workers laid off and the 'leakages' dried up. But debt repayments were so high, net transfers of aid were virtually zero. In 1986 Mobutu kicked over the traces, telling the public: 'You cannot eat austerity.'

Predictably enough, holes began to appear in state funds. Playing the part of sleuth, Louis Goreux, the IMF representative of the day, traced the destination of a missing $100–200 million which had been removed from the export receipts of a state-owned company and sent abroad. 'Somebody had moved the money and that somebody was Mobutu,' recalled Jaycox. Confronted by Jaycox and Goreux, who threatened to suspend lending, Mobutu agreed in 1986 to transfer $20–30 million from his personal accounts abroad to salvage relations. This triggered a remarkable episode which Mobutu must have regarded, understandably enough, as proof positive the international lenders were putty in his hands.

The transfer was made too late for the IMF deadline and, to the annoyance of the officials on the ground who had engineered it, headquarters decided to suspend the programme. A furious Mobutu, who clearly felt he had made a major personal sacrifice, accused the IMF of having lied to him. When it relented and tried to put a new deal together, he told the Fund, to general amazement, to take a running jump. 'One thing the Fund does not like is to be told to go to hell. It was seen as an insult,' remembered Goreux. So a face-saving solution was put together, whereby funds granted under the new programme would quietly be deposited into a special account established on Zaire's behalf in Washington. But Mobutu continued to sulk. Goreux searched around desperately for politicians friendly enough with Mobutu to mollify him. It was only when Jacques Chirac, French Prime Minister at the time, agreed to put in a personal telephone call that the president relented. He could present himself to the Zairean people as the man who had defied the international institutions, while still benefiting from an aid programme.

The image of the Fund going on bended knee to beg one of the world's most corrupt leaders to take its money is not an attractive one. It may help explain why in 1987 David Finch, an Australian economist heading the IMF trade and finance department, resigned over the granting of a new loan, claiming the US had applied undue pressure. The programme staggered along, although it was now a tattered, pitiful scrap of a thing. Kengo had been sacked, and trust in Mobutu's good intentions had shrivelled.

The IMF and World Bank were not the only institutions falling victim to Mobutu's tricks. In 1988, for example, Zaire negotiated a $120 million loan from the African Development Bank. The sum was earmarked for petroleum imports to help the country weather a fuel crisis, and it was granted largely because Cleophas Kamitatu, Finance Minister at the time, was one of the bank's founding members.

Soon after the deal was signed, Kamitatu remembers being summoned by the central bank governor, who told him the president wanted a $40 million cut to cover 'the needs of sovereignty'. Both men's signatures were legally required to authorise the transfer, so when Kamitatu refused, Mobutu telexed over a presidential edict ruling that in future, only the governor's signature would be necessary. Kamitatu was dismissed soon after. 'My successor signed everything,' he wryly recalled.

But the game could not go on for ever. As the economy shrank, the huge bite being taken out by the presidency became more and more painfully obvious. With Mikhail Gorbachev's *perestroika* transforming the Soviet Union, the old Cold War imperatives were fading. Democracy was sweeping across Africa and Mobutu was moving from irreplaceable ally to embarrassment.

In early 1989, another hole in state finances appeared and this time it was too big to ignore: $600–700 million. Jaycox held one last meeting with Mobutu. Previously, their encounters had been conducted before the cameras. Now that relations had turned sour, the president preferred the presidential yacht. This was the only place Mobutu felt safe. Casting off the moorings, he would float midstream, armed guards scouring the horizon, helicopter at the ready.

It was a stand-off Jaycox was scarcely in a state to dominate, having previously caught the tropical disease, giardia. 'I was sick as a dog. I'd been losing weight and had to go to the bathroom every twenty minutes. We were out on that boat and he was making fun of my discomfort. Occasionally he would threaten to throw me to the crocodiles, in a joking way.'

Confronted with the massive financial anomaly, Mobutu's approach—perhaps not surprising given past indulgence—was unapologetic. 'He wanted us to just get over it,' recalled Jaycox with a bemused laugh. 'We were expected to fix it. We documented the gap. He kept talking about his "soldier's word". I indicated that I thought his word was totally worthless—that was the level of discussion we were having—as far as I was concerned his credibility had completely evaporated, the only question now was whether we were going to allow our credibility to follow his down the tubes.'

Mobutu had been given a last warning—but once again he might have been forgiven for assuming his interlocutors were bluffing. In June 1989, US voters elected George Bush, a former CIA chief and long-standing Mobutu supporter, as their new president. The Zairean leader was, amazingly, the first African head of state invited to stroll the lawns with Bush.

The slippages continued. Mobutu was turning sixty, an event he planned to mark by hosting a Francophone summit and major festivities. He was taking the cash he needed for the event from the export receipts of Gécamines, whose restructuring the World Bank was funding. A World Bank letter highlighting these discrepancies triggered an outraged response from Mobutu, who forbade its officials from communicating with Gécamines without government permission. In March 1990, the weary World Bank decided to call it a day and Jerome Chevallier, its resident representative, was asked to act as messenger. 'I went to deliver the letter to Mobutu in the town of Kindu. We'd always had very cordial relations—he would usually address me with the familiar "tu",' he remembered. 'But now he used "vous". "Vous faites du très mauvais travail ici," ("you are doing very bad work here") he told me.' Soon after, Chevallier asked to be transferred, feeling 'anything' could now happen in a country on the verge of an economic precipice.

If the funding of small projects sputtered on, structural adjustment was over, the first time in World Bank history a programme had been suspended with a functioning African government. Even after a series of experiences that could be expected to leave the Bretton Woods institutions allergic to the very mention of Mobutu, there were to be spasmodic attempts to relaunch the aid programme. As foolish an example, surely, of institutional inertia as it is possible to find, they were this time quashed by officials who had, somewhat belatedly, decided to act on Erwin Blumenthal's advice.

Give or take a few doubts about timing, the overriding tenor of my interviews with IMF and World Bank veterans was simple: 'I regret nothing.' Was it really an acceptable answer?

The pragmatic line of argument is that, unappetising as the experience was to prove, Western self-interest made indulging Mobutu worthwhile. 'If we had tried to attach 1990s governance conditionalities to Mobutu, we would have been calling for his overthrow,' says Chester Crocker. 'If we had asked him to turn off the taps, his own people would have toppled him. We would, in effect, have been calling for a *coup*. I'm sure of that.'

But a military putsch in Zaire was only a disastrous prospect if you accepted the premise that there was no better alternative to Mobutu. Many would argue that the West was always overly ready to accept Mobutu's assessment of himself as sole performer on a puzzlingly empty political stage, an impression he conveyed by either scaring his rivals into exile or buying their loyalties. By supporting Mobutu so openly, the West helped bring about that scenario. Sitting in Washington, the economists and politicians never registered how fundamentally their support shored up the domestic image of Mobutu as a kind of malign demi-god, foisted on Zaire by inscrutable alien powers. Dulled since Leopold to the notion of outside forces determining their fate, a defeatist population became convinced he could only be ousted by external intervention.

'When are the Americans going to take Mobutu away?' a parliamentary deputy once asked me as the sun was setting over the People's Palace in Kinshasa at the end of yet another day of pointless wrangling. 'Why can't they come in with their helicopters, like they did in Panama?' The idea that, as a member of the political establishment, it was up to him and his peers to take responsibility for Mobutu's removal had clearly never occurred to him.

Idealists take a different tack, arguing that however faulty the record of Western aid, engaging with Mobutu was a moral obligation. Dictators, supporters of this argument say, thrive in isolation and although much may be stolen on the way, cajoling autocratic leaders to liberalise their economies, open their countries to world trade and set up the institutions associated with accountable government can end up weakening them more dramatically than any amount of foreign disapproval. And what, after all, is the alternative?

'Do we watch a generation of Zaireans go down the drain?' asks Kim Jaycox. 'Is that the smart option? Does that resonate as being wise? Not to me. Looking back, these were good gambles, these were the kind of gambles these institutions were designed to take.'

The problem is that a generation of Zaireans *was* effectively lost, notwithstanding all that goodwill. To cite just a few World Bank statistics, Congo's economy has now shrunk to the level of 1958, while the population has tripled. Average life expectancy is fifty-two, 80 per cent of the population is employed in 'subsistence activities'; illiteracy is growing; AIDS is rife and such diseases as bubonic plague and sleeping sickness are enjoying a vibrant comeback. By the end of the century the government's annual operating budget for what is potentially one of Africa's richest states was dipping below the daily takings of the US superstore Wal-Mart. It is hard to see how, if the World Bank and IMF had boycotted Zaire early on, the situation could have been more disastrous. As debt campaigners in the West point out, the Congolese can now rightly question why they should be asked to repay a penny of the loans made to a man notorious for his dishonesty, which so signally failed to deliver any of the benefits they were promised.

Whether undertaken for hard-headed or high-minded reasons, intervention did no more than fix the country in a kind of purulent agar. True, the country did not fall apart as it had threatened to do after independence. The succession of uprisings, military *coups* and secession attempts that would have probably followed Mobutu's ousting was averted. But it is from just such ghastly experiences that political maturity, inspirational leadership and a sense of direction are eventually born.

Deprived of the chance to learn the lessons of its own history, Zaire's population was kept in a state of infantilism by a more insidious form of colonialism. Instead of the roller-coaster of war, destruction and eventual rebirth, the intervention of the US, France and Belgium, of the World Bank and IMF, locked the society into one slow-motion economic collapse. Balked of expression, unable to advance, mindsets froze over somewhere in the 1960s, leaving the country's leadership at the turn of the century stuck in an ideological time-warp.

CHAPTER TEN

A folly in the jungle

'Anyone wishing to maintain among men the name of liberal is obliged to avoid no attribute of magnificence; so that a prince thus inclined will consume in such acts all his property, and will be compelled in the end, if he wish to maintain the name of liberal, to unduly weigh down his people, and tax them, and do everything he can to get money. This will soon make him odious to his subjects and becoming poor he will be little valued by anyone; thus, with his liberality, having offended many and rewarded few, he is affected by the very first trouble and imperilled by whatever may be the first danger.

'There is nothing wastes so rapidly as liberality, for even whilst you exercise it you lose the power to do so, and so become either poor or despised, or else, in avoiding poverty, rapacious and hated.'

The Prince
—Niccolò Machiavelli

A travel writer of lurid brilliance, Henry Morton Stanley claimed never to have forgotten the horror of his march through the dank forests of eastern Congo, searching for the fabled river he hoped would carry him smoothly through the jungle. 'The trees kept shedding their dew upon us like rain in great round drops,' he recalled in *Through the Dark Continent*. 'Every leaf seemed weeping. Down the boles and branches, creepers and vegetable cords, the moisture trickled and fell on us. Overhead the wide-spreading branches in many interlaced strata, each branch heavy with broad thick leaves, absolutely shut out the daylight. We knew not whether it was a sunshiny day or a dull, foggy, gloomy day; for we marched in a feeble solemn twilight.

'We had certainly seen forests before,' Stanley concluded. 'But this scene was an epoch in our lives ever to be remembered for its bitterness; the gloom enhanced the dismal misery of our life; the slopping moisture, the unhealthy reeking atmosphere, and the monotony of the scenes; nothing but the eternal interlaced branches, the tall aspiring stems, rising from a tangle through which we had to burrow and crawl like wild animals, on hands and feet.'

But if it seemed terrifying to this Briton-turned-American, this verdure, one of the largest expanses of rainforest left in the world, was where Mobutu felt most relaxed. Looking out over the tree tops, an undulating expanse of giant broccoli heads, or driving through the simple villages with their thatched huts, the burden of office seemed less heavy, his spirits quietly lifted and he breathed more easily. He was, after all, a Ngbandi, and this was home.

It was here that, in the late 1970s, Mobutu ordered work to start on a palace. At first this was part of a larger plan to bring development to Gbadolite, the home town he tried, like many an African leader, to transform into a state capital with a simple wave of the presidential wand. But as the years passed and Kinshasa's urban elite showed little inclination to move 700 miles north-east into the depths of the jungle, the president focused his energies on building a citadel fit for a king.

Jumbo jets came and went, ferrying in construction materials, Israeli paratroopers to train the DSP contingent stationed here, hundreds of Chinese workers to build a Chinese village and rare animals, from chimpanzees to Zaire's famous okapi—a curious cross between an antelope, giraffe and zebra—for the private zoo. If he could not force the Big Vegetables to up sticks, he could create a marvel they would discuss over their Kinshasa dinner parties.

Envoys brought Italian marble for a vast mausoleum and chapel, French antiques for the rooms, glassware from Venice. Ironically, the author of authenticity, who had campaigned for the rediscovery of African cultural values, fell for every arriviste cliché in the book. 'I want a marquee for the garden, and I want it now,' the petulant president would tell aides. So the marquee would be brought in by plane, at vast expense.

The airstrip was specially extended to be able to receive Concorde, which Mobutu routinely chartered from Air France and was often to be glimpsed idling on the tarmac. When asked to justify leasing such an expensive plane by a journalist from *Der Spiegel*, Mobutu was unapologetic. 'I cannot sleep at all on a plane and I am terribly scared of sleeping pills,' he explained. 'To accuse me of wasting money—no, I am sorry. Just think of the time I save.'

Initially Gbadolite was a paradise with no Adam and Eve to gambol in it. A restless Mobutu would fly in four or five times a year, his 100-member entourage piling into three aircraft, then touring the grounds in twenty-car convoys. He would stay a few days and be off.

But after Mobutu ceded to domestic and international pressure for political reform in 1990 and announced the end of the one-party state, Gbadolite really came into its own. Mobutu abandoned his base in Kinshasa and spent his time shuttling between his equatorial retreat, the three-storey riverboat moored on the Zaire river and his villas abroad.

The land of his ancestors, Gbadolite was always bound to be of enormous symbolic and spiritual importance to him. But this remote site at Africa's very heart boasted another major attraction, although one rarely mentioned in public. Situated on the Ubangi river, Gbadolite lay only a short hop away from the frontier with Central African Republic, a key selling-point for a man who now had to constantly bear in mind the possibility of an eventual exile. Every hour he spent flying over the rolling forests, away from a capital plagued by protests, was an hour closer to safety. Mobutu left the city a frightened man, having been warned by his generals he ran the risk of assassination if he stayed in Kinshasa. He was also in a state of high dudgeon. In the wake of a brutal security crackdown on Lubumbashi university, staged, awkwardly, a matter of weeks after his groundbreaking political liberalisation speech, foreign allies had started cold-shouldering him. Domestically, the politicians he had showered with riches, nurtured and built up were seizing the opportunity presented by the Sovereign National Conference to turn on their former mentor.

Mobutu always tried not to dwell on his acolytes' hypocrisy. Politicians who denounced him abroad would be welcomed back like prodigal sons. No matter how rude the newspaper article, he never sued. 'He did a lot of forgiving, because there were a lot of betrayals,' said son Nzanga. 'He would say, "Never forget but never take revenge. Because your judgement is not good when you're harbouring hard feelings." ' But the treachery rankled nonetheless.

If Mobutu had read Machiavelli's dictum that it is better for a leader to be feared than loved by his subjects, he had not taken it to heart. Remembering an era when he was fêted in the streets, hailed as the man who had saved Congo from anarchy, Mobutu could not get used to being hated. 'Something died in him from that moment on,' judged his closest former aide, Honoré Ngbanda. He noted how the president would interrupt important audiences aboard his yacht to rush on deck and acknowledge the chants of praise from passengers on passing ships, so desperate had he become for applause.

In a fury of 'I'll show them', he decided to leave his ungrateful countrymen to the multi-party democracy they wanted, taking bitter satisfaction from the ease with which he sabotaged the process with the gift of a Mercedes here, a bank deposit there. His revenge was to live in style. Nothing pleased him more than to invite a group of Western VIPs up to what had been nicknamed 'Versailles in the Jungle' by the press and watch them gawp. 'It was so incongruous your mouth fell open,' remembered one regular US visitor. 'There was this big golden pagoda at the airport. The reception area was so enormous if someone was sitting on the other side you couldn't recognise them. Then you'd drive past thatched huts and untouched, indigenous villages and there would be this palace like the Louvre. It was indescribable.'

The VIPs came, duly gasped over the musical fountains, the swans gliding over ornamental lakes, the model farm with its 500 Argentine sheep, and Mobutu's heart swelled with proprietorial pride. But he was blind to the true nature of their amazement. It was not astonishment at a job well done. Mingled with the patronising contempt of the Old World sophisticate for the tackiness of the nouveau riche was shock at Mobutu's insensitivity, disbelief that the leader of a country in such desperate straits should have permitted himself such extravagance without registering the message such crassness carried to the outside world, and, underlying it all, the horrified realisation that this was where large amounts of Western aid had ended up.

For Gbadolite was the ultimate in African presidential follies. There was just too much of everything: too much champagne, too much beer, too much marble, too much gilt. It was Mobutu's Graceland, a fittingly vulgar monument to a vast ego, part of the answer to the question of just how a single—albeit extended—family could manage to consume quite so much of a country's wealth.

So it was poetic justice that Gbadolite should eventually prove a folly in quite a different sense. Mobutu's move there—a gesture of petulance towards a population that had turned against him, a barely disguised appeal for love—was to be his eventual undoing, the worst mistake of a career until then characterised by a superb instinct for self-preservation.

Like it or not, every great man is doomed to acquire a dragoman, some lesser mortal appointing himself as interpreter and guide. Intermediary between the world and his boss, he hopes, like some pilot fish leading a Great White shark to its prey, to grow fat on the morsels trailing from the kill. In Mobutu's case, the position was always bitterly contested, but for one brief moment a young white businessman was naive enough to think it was his for the taking. Pierre Janssen became a member of Mobutu's court, the only European to join the president's intimate family circle, thanks to a fortuitous meeting in the Hilton hotel in Brussels.

A young businessman in a hurry, proud of having worked out early on in life that 'money creates power', he was introduced by a Zairean friend to Yakpwa Mobutu, Mobutu's daughter by his first wife. It was his natural predilection for black women, rather than his fascination with celebrity, he claims, that drew him to 'Yaki', as she is known. They chatted, found each other mutually attractive and a courtship began. Two years later they were married.

Janssen is interesting because, as a newcomer to the Mobutu world, he did not share the blasé vision of the Zairean elite that tagged alongside the president as he jetted around the world. To the son-in-law, it was all new and amazing. His impressions—the excitable commentary of a Belgian social climber sensing untold riches inching within his reach—are those of Everyman, pressing his nose against the windows of the rich and famous.

He is no James Boswell, dutifully recording the *bons mots* and thoughtful musings of his master. The book Janssen wrote as a result of his experiences contains florid accounts of voodoo sessions in the Mobutu household and melodramatic descriptions of secretive meetings with freemasons. It also vaunts an intimacy with the president which is challenged by members of the family. 'You can take that book and put it straight in the bin,' Mobutu's son Nzanga told me with distaste. Yet it contains insights that only a man with the most materialistic of fantasies could contribute.

Janssen has the mind of a grocer. He clearly expended a lot of energy during his time in the Mobutu household making mental estimates: how much Mobutu spent on champagne, how much on cars, how much on jewellery, how badly the restaurant overcharged him, how blatantly his aides stole. The result, in his book, is a shopaholic's catalogue, an account that sheds fascinating light on the minutiae of a kleptocracy, the lifestyle a former cook's son had come to regard as normal after three decades in power.

The son-in-law's fascination with Mobutu had been heightened by a fall from grace mirroring the president's own. At thirty-five, he presented himself when we met in Paris as a man abandoned by fortune, paying a high price for his well-meaning involvement in Zaire. Now separated from Yaki, he was no longer on speaking terms with his African in-laws, who regarded him as little more than a gigolo. The French publishers of his memoirs had gone bankrupt, his business ventures had crumbled to nothing, the Cap Ferrat house where he was staying did not, he promised, belong to him. 'I'm ruined, I'm on the street,' he said with a bitter laugh. 'When I went to Kinshasa I had my own career, I earned a good living. Now I'm separated from my wife, I have "Mobutu" stamped on my forehead and I can no longer go back to Congo. My wedding was the worst day of my life.'

Yet his fleshy, sun-kissed face hardly spelled deprivation. And he had the cocktail-goers' habit of avoiding eye contact, constantly scouring the expensive Chinese restaurant we had retired to for someone more interesting to talk to. As his search was rewarded ('Look, there's John Galliano'), I realised Janssen, who confesses in his book that he always travelled first class because it increased the likelihood of a brush with a VIP, had probably picked the spot on the elegant Avenue Montaigne precisely for its guaranteed celebrity quotient.

There were certainly plenty of VIPs on offer when he married Yaki in Gbadolite on 4 July 1992. The 2,500-strong guest list included regional presidents, Saudi princes, Middle Eastern dignitaries, foreign ambassadors and the entire Zairean government, although not, to Janssen's disappointment, members of Monaco's royal family. A chartered DC10 and two Boeings had shuttled between Europe and Zaire to muster them in Gbadolite. It was one of the rare occasions when Mobutu could put the $100 million complex to good use. The president had gone a little over the top in designing the main palace at Gbadolite. Sprawling across 15,000 square metres, its seven-metre malachite doors were so heavy it took more than one man to open them—this was a building designed for giants. The huge marble-lined salons were impossible to fill.

Belatedly, Mobutu realised that he could not live with such grandeur and ordered a second palace on more human scale to be built at Kawele, a few kilometres away, complete with discotheque, Olympic-sized swimming pool and nuclear shelter. With its Louis XIV furniture, Murano chandeliers, Aubusson tapestries, monogrammed silver cutlery and walls hung with green silk—green was Mobutu's favourite colour—Kawele was hardly a hovel. But it was positively cosy compared to the main monstrosity, unused except for special occasions such as his daughter's wedding.

For the ceremony, the bride wore haute couture: a hand-embroidered Jean Louis Scherrer wedding dress with a six-metre train, costing $70,000. Later, she donned a Nina Ricci salmon-pink outfit with silk trimmings. Throughout the day she alternated the three gem clusters bought from the jewellers of Paris's Place Vendôme, a wedding present from her father, Janssen estimated, worth a total of $3 million.

After the religious ceremony the guests, wilting in the equatorial humidity, moved to the reception, where a meal of lobster, salmon and caviar awaited, washed down with around a thousand Grand Cru wines from Mobutu's 15,000-bottle wine cellar. There was a massive firework display and three orchestras provided live music. But the *tour de force* was the wedding cake, a four-metre concoction of chilled meringue and cream in danger of melting in the tropical heat. Prepared by a Paris chef that very morning, it had been dismantled, loaded in pieces onto a special charter, and flown to Gbadolite: a fourteen-hour round trip costing, estimated Janssen, $65,000.

The honeymoon? A holiday on the Thai beach resort of Phuket, where a local king played host. And a life of ease awaited the couple on their return. Mobutu had been generous with his gifts: a villa in Uccle, the chic district of Brussels, another in Kinshasa, an envelope stuffed with $300,000 in cash and the promise of an apartment in Monaco.

With an introduction such as this, it would have been almost impossible for a man like Janssen not to aspire to more. Although he states he was the only member of Mobutu's entourage to turn down regularly offered cash presents, Janssen clearly expected his marriage to open up all sorts of attractive business opportunities. Despite repeated warnings from his more sceptical wife not to meddle in what he did not understand, he set about cosying up to his new father-in-law.

First Janssen tried to take over the running expenses of the Mobutu household. Then he proposed the president name him Zairean consul to Monaco, a title, he believed, that would put him in an ideal position to court the rich Arabs holidaying in the principality. There was a plan to revitalise Zaire's failing palm oil industry, another which involved helping Libya break UN sanctions. One by one, the plans were hatched, matured and, as often as not, according to Janssen, torpedoed in a subtle whispering campaign mounted against him by jealous in-laws and suspicious aides who had nabbed the post of dragoman long before his bumbling arrival on the scene.

In the process, however, Janssen's image of Mobutu slowly changed. 'I'd built up an image of a terrible dictator who killed people,' recalled Janssen. 'The man I came to know bore no resemblance to that. I saw a man who was very sensitive, a very good head of the family, a man who loved his children above everything and loved his country, but had weaknesses, like everyone else.'

Mobutu's day began at half-past six. At seven a team of masseurs from the Chinese village would knock at his door to give him his daily work-over. At eight, after reading the international press, he would eat breakfast on the terrace, throwing the odd crumb to the peacocks wandering through the formal gardens. At nine he entered his study and the first of what would be a series of bottles of Laurent Perrier pink champagne would be uncorked.

Lunch was often that Belgian speciality, moules-frites, with the mussels flown in from Zeebrugge, washed down with a 1930 vintage, in tribute to the year of Mobutu's birth. When a room needed brightening up, flowers would be flown in from Amsterdam. A barber from New York, a hairdresser from Paris, the French couturier Francesco Smalto: they were all summoned and flown in across the continents whenever deemed necessary. 'They chartered Boeings like most people use supermarket trolleys,' recalled Janssen.

To unwind, Mobutu listened to Gregorian chants, a taste he may have acquired during the years spent being educated by Belgian priests. But there were few occasions for that, given the constant stream of visitors. From Kawele, perched on a hill with clear vistas around, both Mobutu and his DSP guards were perfectly positioned to check out arriving visitors before they rolled up at the house. If it was someone special, Mobutu might call for his personal Chevrolet and drive himself and his guests off fishing or to some quiet forest glade for a picnic, washed down with champagne from a monogrammed silver ice-bucket. This could be a hair-raising experience, as Mobutu was not a good driver. One former State Department official, a frequent visitor to Gbadolite, was surprised to find himself hanging on for dear life as Mobutu and his clearly petrified bodyguards careered down the track, sending pedestrians flying into the bushes. 'It was like a cartoon, people and things were leaping out of the way. The ambassador made some remark and Mobutu said: "It's OK, these are all *my* roads." '

Despite the distance from Kinshasa, the queue of supplicants waiting patiently for his attention was endless. Most, whether opposition or MPR politicians, family members, foreign visitors, came in search of one thing—one of the hefty envelopes kept in Mobutu's desk drawer stuffed with $100 bills. 'He paid out, and paid out. He was surrounded by leeches thirsting for dollars,' recalled Janssen. 'I looked into his eyes and I felt sorry for him.'

In every man's life, the same events, the same characteristics can be viewed by his intimates from radically different points of view, perspectives so far apart the final picture may be unrecognisable to rivals claiming exclusive insights. So it was with Mobutu—always at the heart of a fierce 'war of influences'. For the president's sons and daughters, the years in Gbadolite were a time when Mobutu, estranged for too many years by the affairs of state, developed a new set of priorities and took a well-merited break. The loss of two sons—Niwa and Konga—to illness had given him a sharp appreciation of the fragility of human relations, the importance of family. 'We saw him a lot more than we ever did before,' remembered his son Nzanga. 'He tasted joys he had never known, he rediscovered family life.'

To outsiders like Janssen, the same scenario appears in a different light. For them, Mobutu had fallen victim to a predatory family which, exploiting his desire to be a good paterfamilias, was proving as voracious in its demands on Mobutu as he had been in his demands on the state. It is impossible, looking through their eyes, not to pity a man whose personal relations had become, after decades of patronage and bribery, utterly corrupted by the issue of what he was in a position to give.

Despite his ready generosity—or rather, because of it—Mobutu was hardly getting value for money, Janssen claimed. Talking to the cook at Gbadolite, he discovered that Mobutu was paying three times the wholesale price for the 10,000–12,000 bottles of champagne the household got through each year. In Cap Ferrat, Mobutu paid twice the going rate for the fleet of Mercedes he rented. In Brussels, well-heeled Zairean exiles would actually drive to Mobutu's residence to talk for hours on the presidential satellite telephone, knowing the bill would be settled without question. He was being ripped off by everyone, from the Belgian steward who was finally sacked when Mobutu twigged to his systematic overcharging, to the ambassadors who arranged his lodging on foreign trips and the Senegalese marabouts (witch-doctors) consulted over every major decision. Trusted implicitly by Mobutu, who effortlessly juggled a belief in the African world of spirits with the Catholic faith, they were paid twice—once by the president, a second time by the politicians who had asked them to guide the president in any given direction.

The worst offenders, Janssen asserted, were often his own children. He claimed one son, dispatched to the United States to buy six armoured Cadillacs of a type used by George Bush, inflated their price by $40,000 apiece, then telephoned from New York claiming he had lost the $600,000 Mobutu had given him during a mugging. His trusting father wired another $600,000 to New York only to see his son return, empty-handed, claiming the model was no longer available. The 1993 riots, he says, provided the same son with a lucrative sideline: borrowing the presidential yacht *Kamanyola*, he ferried fleeing mouvanciers across the river to Brazzaville for $2,000 a family. 'Everyone was stealing from him, exploiting his inexhaustible generosity. And Mobutu appeared not to notice,' claimed Janssen.

For many observers, such excesses could be traced back to the death of Marie Antoinette, Mobutu's first wife, who had expired of heart failure in 1977. Still remembered with fondness by the populace, she had been a restraining influence on Mobutu, bringing out the best qualities in his character, holding his vices in check. 'When Marie Antoinette was alive she acted as a kind of barrier against the demands of the clan, even though she was a Ngbandi herself,' claimed a childhood acquaintance. 'When she died there was a general slackening. The clan began to take over everything.'

Mobutu's extraordinary personal arrangements merely added to the problem. Marie Antoinette's place was taken by Bobi Ladawa, a former mistress. But despite the many dalliances tracked by Kinshasa's embassies, Mobutu still felt the need for what the Congolese coyly refer to as a 'second office'. While many married men seek relationships with women who differ radically from their wives in appearance or character, Mobutu opted for the truly familiar. Kossia, none other than Bobi Ladawa's identical twin, became his mistress. Many Zaireans, spooked by the fact they could never tell which woman, wife or mistress, was perched on Mobutu's arm during official occasions, believe the arrangement represented a good luck charm for superstitious Mobutu, for twins are regarded in many parts of Africa as possessing totemic significance, a mirror-image combination blessed with magical powers. 'It was a way of warding off his first wife's angry spirit,' one Congolese official explained. 'With a twin on each side Marie Antoinette couldn't get at him.'

Another possibility is that the president felt compelled to bed the sister to avoid being cuckolded, as whoever married Bobi Ladawa's twin would, in a way, be savouring intercourse with the first lady. The custom of appropriating your wife's sister is, in any case, practised in the equatorial region. 'You always go upwards, never downwards,' my driver François once explained to me. 'You can sleep with your wife's elder sister, but not her younger sister.' For all but the midwife present at the birth a twin, presumably, counted as neither up nor down.

On the surface, it was a fairly cosy *ménage à trois*. One diamond trader invited to dinner in Gbadolite remembered joining Mobutu at table with the two women—once strikingly beautiful, their features now blurred by envelopes of fat—sitting on either side, discussing housekeeping arrangements with no visible tension in the air. But the sisters inevitably competed for attention, joining forces when necessary to champion the interests of Bobi Ladawa's children against the rival claims of Marie Antoinette's offspring. 'There was a problem of rivalry between them, but they were intelligent women and at a certain point they realised they might as well unite their forces,' was Janssen's view.

So the president was surrounded by family members vying for his love, and in the Mobutu household, of course, terms of endearment were expected to be expressed in strictly financial form. In any case, how could a man who had turned embezzlement into a presidential art now lecture his family and servants on honesty?

There was another factor preventing Mobutu from cleaning out his own Augean stables. Paradoxically for a president branded as one of the greediest heads of state of all time, his lack of understanding of the workings of an economy was only matched by the absence of any grasp over his own domestic expenses.

Shielded for decades from the practicalities of daily life, he no longer knew what items cost in the real world or what normal people spent on lodging and food, floundering disastrously once when a journalist asked him during an interview what the price of bread was in Kinshasa. 'For such a big thief he was rather naive about money,' mused a diplomat.

The purse strings were firmly in the hands of the twins and their own lack of sophistication—the simplicity of the African peasant who keeps her money tucked under her mat and never goes near a bank—was manifest in their preference for cash. Cheques and credit cards appear to have been almost unheard of in the Gbadolite household, where debts were settled and presents made with piles of banknotes. Before each trip, $100 bills would be packed, gangster-style, in Louis Vuitton briefcases, which would be lined up in the hall in descending order of size.

The president had no idea of daily incomings or outgoings. According to Janssen, 'Mobutu never checked a bill. He would say to his aides: "I'm going to Switzerland tomorrow, get me three million dollars." But where that money came from, he just didn't care. It was up to them to find it.'

Nevertheless, the huge drain on his resources represented by the palaces and their denizens was not to be the cause of Mobutu's downfall. By withdrawing to his jungle hideout, he had made a mockery of the post of president, rendering himself increasingly irrelevant to unfolding events. Theoretically, a well-equipped rebel movement would only have had to bomb the airstrip at Gbadolite and jam his satellite telephone signal to render him utterly redundant. But Mobutu was carrying out the task of progressive isolation voluntarily.

Insidiously, his decade-long sulk was depriving him of the first-hand contact he needed to gauge the mood of the country, the public and his armed forces.

For prime ministers, ambassadors and visiting foreign envoys, any presidential ruling now involved a four-hour round trip into the heart of the jungle and sometimes days spent waiting for an audience. In a nation which had become the repository for the world's obsolete aircraft, the trip could prove fatal. A foreign minister, a Tunisian ambassador and a contingent of soldiers were amongst those who paid with their lives for a Gbadolite visit when their jets crashed approaching Kinshasa's Ndjili airport.

Jack Lunzer, a British diamond dealer, narrowly escaped a similar fate during a flight that underlined how marooned Mobutu was becoming. Summoned to Gbadolite, he became alarmed as the hours passed with no sign of imminent descent. The Belgian pilot then confessed that he was getting no navigational signals from the Gbadolite control tower, which appeared to have run out of fuel. With the cloud cover low over the trees, landing unaided was out of the question and the plane had only 50 minutes worth of fuel for the one-hour trip to the nearest Zairean airstrip. Slowing the engines to a crawl, the pilot nursed the aircraft back to the town of Mbandaka as his passengers counted the minutes. 'As we touched the runway, the engines went splutter, splutter, splutter and stopped. When we got out, I was dripping with sweat,' said Lunzer.

Once in Gbadolite, the president's attention had to be captured. Increasingly, visitors found planned tête-à-têtes hijacked as the court headed off on one of the country jaunts Mobutu adored. Honoré Ngbanda, the president's closest political aide, recalled the difficulty of focusing the president's mind on such pressing problems as the refugee crisis in eastern Zaire and the burgeoning Kabila rebellion while his boss insisted on playing the part of country squire. 'The President of the Republic no longer had an office! We would meet in farms on the outskirts of Gbadolite, in the middle of fields of maize and manioc, amid the commotion of farm machinery and labourers' cries. It was difficult to discuss urgent and sensitive issues in depth.

Even if his capacity to assimilate dossiers and issue orders was legendary, this method of working was no longer appropriate given the seriousness of the crisis facing the country.'

Mobutu's horizons, visitors noticed, were shrinking with every passing year. Like the French royal courtiers in Versailles who played at being shepherds and dairy maids as the shadow of the guillotine crept closer, Mobutu had developed a passion for glad-handing local farmers, talking about crops, soils and rains.

The agricultural interest had initially seemed contrived. Andreas Wagner, a Swiss vet hired by Mobutu, accompanied forty high-yielding dairy cattle on a C130 flight that took them straight from Switzerland to Gbadolite in 1979. Struggling with a desperate shortage of grazing, Wagner, who had been told this was a project aimed at bringing agricultural development to the area, swiftly realised he was working on a cosmetic showcase, the result of a presidential whim. 'Mobutu came once for ten minutes to inspect the cattle and I tried to explain the various problems. But you can quickly tell when someone isn't listening. He wasn't interested at all,' said Wagner, who left in a hurry when he discovered his complaints had triggered the issuing of an arrest warrant against him.

Yet, by the 1990s Mobutu had become more engaged. This was where he had started out, running barefoot through the fields, helping his mother plant cassava, living off the land. He was rediscovering both his tribal roots and the simple pleasures of childhood. He loved being able to reach up from his open-backed car and pick fruit from a tree planted under his instructions. At times, he wondered if he had missed his vocation. 'He'd always say—"if I could do it all again I'd be a farmer",' said Nzanga. 'I think that is what he really would have liked.'

Daniel Simpson, the US ambassador during those last years, once drove out to Goroma, the model farm Mobutu had planted 10 kilometres from Gbadolite, for talks. 'He'd picked up cuttings in Egypt and on his various trips and he and his wife were personally supervising the planting of orange and lemon trees.' As Mobutu passed, Gbadolite's residents lined the route to cheer the man who had brought them electricity and telephones, just as Kinshasa's population had done in the good old days, before they turned ungrateful. Increasingly, Mobutu spent his days outdoors, his attention focused on clan disputes, village problems, handing out cash presents. If he had failed in the task of ruling his enormous country, he could still play the part of tribal chief with conviction.

It was all utterly maddening to top-ranking visitors needing decisions and it was made more infuriating by the 'Uncle Fangbi' factor. Just as Ngbanda or Vundwawe Te Pemako, Mobutu's other key adviser, were broaching some subject of enormous sensitivity, Uncle Fangbi, the brother-in-law, would interrupt to remind the president his siesta was due, or that he had promised to drive to a nearby village.

A former nurse's assistant, Uncle Fangbi knew little of politics or military matters. But in the final years the man who had dubbed himself 'Representative of the Presidential Couple' had enlarged his field of influence so that it expanded from responsibility for Mobutu's daily schedule to approval of his private and professional audiences, control of the household budget and, finally, the power to suspend officials working in Gbadolite.

Uncle Fangbi would arrange outings without revealing the destination until five minutes before departure, making security checks impossible. He allowed Mobutu to drive the first car in the convoy and encouraged the president, whose paranoia jostled with a desperate desire for some spontaneity, some freedom in his life, to go village walkabout and sample the local palm wine. The DSP hated Uncle Fangbi not only for overruling them, they claimed he was pocketing their salaries. But when aides remonstrated, Mobutu would say 'I'll see to it'—his way of signalling a conversation was closed.

The president's fantasy appeared to be a quiet old age, culminating in a deathbed scene surrounded by family and adoring dependants. He had built a marble mausoleum which held Marie Antoinette's body, and his own name and those of his children had been carved into the stone in expectation of their eventual demise. Despite all the precautions, despite all the investments abroad,

Mobutu hoped never to leave his country. 'He never anticipated a life in exile,' said Janssen. 'He used to tell his aides: "I will never read the *Cap Martin* (one of the Riviera newspapers)." '

Did Mobutu not realise the risk he was running by indulging in such escapism? Today, each aide, each politician, each family member will tell you how many times he warned Mobutu, how often he sought to persuade the president to return to Kinshasa, how thoroughly it suited various ill-intentioned players for Mobutu to marginalise himself. 'My conscience is clear. He was listening to Vundwawe and Ngbanda at the time,' insisted Nzanga, the son. 'I know it suited them for him to stay away from Kinshasa. They came to Gbadolite and made my father sign decrees, name heads of enterprises. All that was to their benefit.' Not so, claimed Ngbanda. 'We tried to bring him back to Kinshasa, but without success. He had been engulfed by the family.'

As the power vacuum widened, the refrain was even taken up by increasingly anxious foreign heads of state—to no avail. 'I'm waiting for the elections. If I win, I'll return to Kinshasa. If I lose, I'll stay here,' Mobutu would growl in response. The man who had entranced the crowds must have known better than anyone the importance of contact with the public, a constant monitoring of the nation's temperature. If intellectually he registered the danger he was running, it is clear that he simply could not bring himself to do anything about it. Like Mr Kurtz, he had been swallowed up by the forest and the fantasies it spawned.

One by one, he was calling a halt to the duties that had formed the framework of his working existence. 'At one time Mobutu would personally swear in all army officers. They would swear allegiance to him, eyeball to eyeball,' recalled Daniel Simpson. 'By 1986 he'd stopped doing that. Another thing that stopped was his travelling around the country. His officials drew up a schedule of trips and tours for the electoral campaign and he said "forget it". It just didn't give him any pleasure any more.' The arch manipulator appeared to have lost his taste for politics, hardly able to summon a flicker of interest even in the workings of government that had once obsessed him. 'At first he used to sit in on every cabinet meeting. But then he became less and less interested. It became once a month and then he practically didn't chair any at all,' said a former prime minister.

Who can blame him for feeling jaded? One senses a colossal fatigue with leadership and the burdens of office, the weary boredom Larry Devlin first caught a glimpse of when, returning to Kinshasa in the 1970s, he found his old ally drowning in sycophancy. As the former CIA chief put it: 'Having accepted the fact that he was a genius, what more could he do?' He had seen human nature at its least inspiring: grasping, insincere and treacherous. He had learned, like some emissary from Hades sent to test humanity's depravity, how to encourage and exploit those very qualities.

Despite all the money and attention lavished on it, the Gbadolite complex never seemed a very permanent fixture. Waiting once for an audience with Mobutu, Leo Tindemanns, a frequent visitor, noticed the monkeys scampering over the grounds. 'I was struck by the fact that once Mobutu left the place would return very rapidly to the bush,' said the former Belgian prime minister.

So it proved. When Kabila's forces seized the area the site was trashed by Mobutu's loyal villagers and rebels alike. They emptied the wine cellars, barbecued the imported herds, drove off with the fleets of Mercedes and stripped the palaces of furniture and fittings, which ended up gracing market stalls in Bangui, capital of Central African Republic.

The walls once hung with green silk are scrawled with graffiti now, weeds have sprouted in the empty swimming pool and broken glass crunches underfoot in the salons where Janssen celebrated his short-lived union to Yaki. The chandeliers have been used as target practice by bored soldiers. Only the Chinese palace, with its ornamental ponds and dragon statues, has been left structurally intact, along with the monstrous marble baths, too big even for the most determined looter.

Kabila's Tourist Ministry played briefly with the idea of turning

Gbadolite, and the yacht *Kamanyola*, into a destination for travellers with a taste for the unusual, stops on a guided tour entitled, perhaps, 'Lifestyles of the Rich and Infamous'. A group of French businessmen even flew to Gbadolite to assess possibilities, but were discouraged by how few of the original fittings were left.

Tourism had to be put on hold, in any case, when a new war broke out in 1998 and an army unit from Chad, one of the African countries that took Kabila's side in this second conflict, set up headquarters in Gbadolite. When the Chadians tired of their foreign military adventure, rebel forces led by Jean Pierre Bemba, the son of one of Mobutu's businessman friends, seized control of the area. The Ugandan soldiers backing Bemba's campaign moved into the palace at Kawele, after first reinterring the bodies of family members dragged—in just the nightmare scenario Mobutu always dreaded—from the marble mausoleum. 'There are limits to revenge,' explained one of the rebels.

Bemba has been hailed as a liberator by a hungry local population, desperate to see a return of the investment and jobs of the Mobutu era. But Gbadolite's moment in the sun has passed. A future president is unlikely to be a Ngbandi, and only the most self-assured successor would be so foolhardy as to risk retiring to Gbadolite. Like many another presidential home town suddenly deprived of the protection of its local hero, the region Mobutu was so determined to help is now doomed to return to the obscurity from which it briefly emerged.

Like the Swiss cows and Argentine sheep that slowly wasted in the tropical heat, Gbadolite was an artificial construct, wholly dependent on Mobutu for its existence, incapable of surviving his demise. And that had become inescapable. For while the president picnicked under the trees, admiring the scenery and bonding with his relatives, younger men with their own voracious extended families to satisfy were plotting and scheming. His failure to heed the constant warnings ensured that when he finally left Zaire he did so not with the dignity appropriate to one of Africa's longest serving leaders, but in a sweat-soaked flurry of terror, humiliation and betrayal.

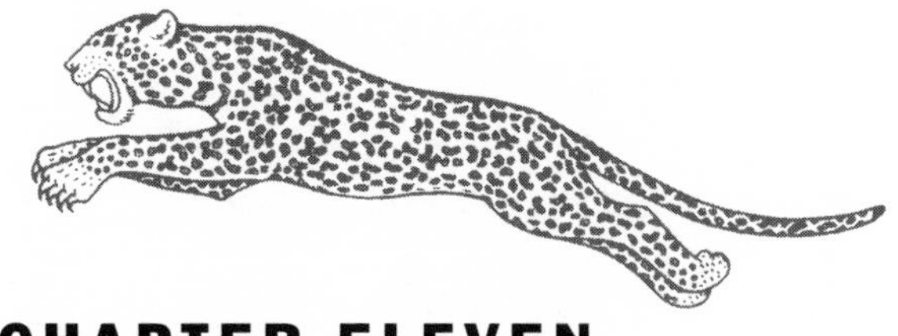

CHAPTER ELEVEN

The night the pink champagne went flat

'It's when it rains that you can pee your pants with a quiet mind.'

Kasaian proverb

Just as every baby boomer knows what they were doing when they heard that John Kennedy was shot, I can pinpoint where I was on 6 April 1994—a day that was to mark a generation of Africans as deeply as the Dallas shooting marked Americans.

I had been invited, along with a handful of Zairean journalists, to a late-night meeting with the then head of the country's customs office. Thanks to the huge opportunities his job offered him for bribe-taking, he was said to be one of the richest men in Zaire, richer even than Mobutu, who allowed him such leeway because he regarded this man, who had once tutored his children, as a kind of adopted son.

One of the generation of what had been dubbed 'the baby dinosaurs', he was inordinately vain, and his ballooning wealth had made him ambitious for a role in politics. He was now gunning for the prime minister's job. No one who knew the political scene believed he could get it, but for some inexplicable reason he seemed to think that getting the press to talk about him would convince the old man of the value of his candidacy.

Sporting the Sunday best appropriate for such a social event, we wound our way up through the streets of Binza, home of the Kinshasa elite, and hooted our way into the courtyard of a new house. It was the usual white-washed, heavily guarded Zairean pile, although, as building was still in progress, the décor had not yet reached its full, tacky potential. I sneaked a visit to the bathroom and confirmed, to my enormous satisfaction, that it would have gold bathtaps.

Sitting on the balcony overlooking Kinshasa, with the lights of Brazzaville in the distance, we listened politely as our host boasted, with all the subtlety of a second-hand car salesman, of his political acuteness. He had the president's ear, we were to understand, knew so much but could, unfortunately, tell us so little. His wife was not in evidence. Instead, playing hostess was a sultry female presenter who read the television news, the ultimate rich man's trophy.

While we could hear the sounds of dependants tucking into a hefty meal inside, on the balcony the diet was liquid—Mobutu's beloved pink champagne. Leave your glass untouched for more than a couple of minutes and our host would throw the contents ostentatiously over the balcony and pour a top-up, exclaiming: 'You can't drink this, it's lost its fizz.' Staged for the benefit of guests who probably earned less in a month than he was tossing into the night sky, it was a crudely effective display of wealth.

We were all fairly muzzy by the time the Telecel rang. There was a brief exchange in Lingala and when our host put the mobile phone down his eyes were big. A plane carrying Rwanda's and Burundi's presidents had been shot down coming in to land at Kigali airport in Rwanda. Mobutu, it was said, had been planning to take the same flight but had changed his plans at the last moment. Both leaders were dead.

There was a baffled silence. Who could have done it? What did it mean for Africa, and Zaire in particular? The party broke up, murmuring unanswered questions.

The consequences were to be cataclysmic. And it was appropriate that I heard the news on that balcony, with a man who epitomised all that was worst about the Mobutu regime. For the downing of that distant presidential jet in a tiny hilly country half a continent away represented the toppling of the first in a row of dominoes stretching

1,000 miles, all the way from the cool hills of Rwanda's capital to the torpid heat of Kinshasa.

Mobutu spent that night in tears, mourning Rwandan President Juvenal Habyarimana, a personal friend, fearful for the future. He was right to weep. The tremors set in motion on 6 April were to bring down his regime three years later, lowering the curtain on Cold War politics in Africa and marking the continent's entry into new, uncharted waters.

The dead Habyarimana was probably offered up as sacrificial lamb by the extremists in his own Hutu community. Angered by their leader's attempts to hammer out a power-sharing agreement with the Rwandese Patriotic Front, a rebel group dominated by the minority Tutsi ethnic group, the fanatics in one fell swoop rid themselves of a supposed traitor and gave the Hutus an excuse for a ruthless crack-down on the Tutsi minority living among them.

Long-nursed plans for the massacre of the Tutsi community that had once constituted Rwanda's aristocracy were put into effect by local officials who counted on the instincts of unquestioning obedience nurtured in one of Africa's most rigidly bureaucratic states. Sure enough, Hutu villagers did precisely as they were told. With the militias known as interahamwe—'those who stand together'—leading the way, they turned on their Tutsi neighbours. Within three months Rwanda was littered with piles of stinking bodies. Between 500,000 and one million Tutsis and Hutu moderates died in the world's quickest genocide, much of it carried out with that most basic of killing instruments: the machete.

The massacres had the opposite effect of what was intended. The Hutu extremists were aiming for eventual control of a mono-ethnic state. Instead the RPF, whose fighters the Hutus dubbed 'the cockroaches', stepped up their military campaign. By July it had won control of Kigali and the Hutu extremists had fled into neighbouring countries. Warning of certain Tutsi revenge for a slaughter condoned by an entire community, they took with them over two million peasants. Laden with straw sleeping mats and cooking pots—the bare essentials of existence—more than a quarter of the Rwandan population abandoned their villages. It was the largest, most sudden human exodus in modern history.

Stripping the landscape as they passed, a human swarm that gobbled up woods, livestock and crops, the Hutus headed for the borders with Tanzania, Burundi, Uganda and Zaire. Hour after hour, hundreds of thousands of bare feet scuffled and scurried through overwhelmed crossings, sending up a whispering chorus of guilt and fear. Then, the frontier safely passed, the refugees stopped. More than half ended up in Zaire's Kivu region, settling on the unforgiving black rock of Goma, Bukavu and Uvira.

At first, when a cholera epidemic felled tens of thousands of the refugees, they were viewed by the West as helpless victims of an ethnic conflict. One of the most complex humanitarian operations the world had ever seen got underway in Africa's Great Lakes region, with 200 aid organisations bringing in medicine and shelter, doctors and nurses, food and water. As the immediate crisis passed, a rather more sinister status quo began to emerge from the soft grey blanket of mist that formed over each settlement, product of innumerable charcoal fires.

Encouraged by the relief organisations, who found it easier to distribute aid through recognised chains of command, the mayors and prefects who had masterminded Rwanda's genocide neatly re-established control over their communities, with the interahamwe and army soldiers providing the muscle to police a government-in-waiting. The men who featured on the lists of human rights organisations investigating Rwanda's genocide had not been sidelined by the community they had so sorely misled. Instead, they decided who got fed, how much, and even levied a form of tax. Determined to prevent a mass return which would deprive them of their constituency, they told the gullible they would have their eyes plucked out if they returned to Rwanda. The bodies of those who dared to defy them would be found by aid workers in the morning, a blunt lesson to the rest.

Like a monstrous cancer, the camps coalesced, solidified and implanted themselves in the flesh of east Zaire. An exile initially expected to last a few weeks turned into months, then years. Time and time again, the office of the UN High Commissioner for Refugees (UNHCR) would announce that conditions were ripe for a mass return. Transport was laid on, way stations prepared, the supposed support of community leaders secured. The buses would leave virtually empty, their handful of passengers drawing silent stares from the crowd. Any return, the exiled Hutu extremists had decided, would be led by a conquering army. To that end, the Rwandan former generals and militiamen were rearming and recruiting, making a mockery of the camp dwellers' supposed refugee status. So confident were the extremists, they even trained young fighters within sight of passing aid workers. Preoccupied with their humanitarian targets, aware the international community was not ready to tackle the huge problem posed by the hardliners, the relief officials looked the other way.

Increasingly, there seemed little reason to move. Seen from the air, from where the alleys, distribution points, clinics and individual prefectures dividing up these mosaics of blue, red and green tarpaulins made sense, it was clear these were towns rather than camps, blessed with all the to-and-fro, the ceaseless commercial activity, of any sophisticated urban conglomeration. When it came to adapting to adversity, the Rwandan refugees could teach even their inventive Zairean neighbours a thing or two about Article 15. The cattle herds that were the source of Goma's famous cheeses slowly disappeared from surrounding hills, rustled by Rwandans, who operated their own camp abattoir. Meat in the camps was so plentiful, it was actually cheaper than in central Goma. Local wildlife—from flayed monkey to chunks of hippo—provided an exotic alternative. Penetrating the nearby Virunga National Park, a former tourist attraction, refugees took whatever came in handy. The denuded areas left as they felled woodland for charcoal were so large, they were visible on satellite photos.

In 1995, a UNHCR survey listed nearly 82,000 thriving enterprises in the camps, including 2,324 bars, 450 restaurants, 589 general shops, 62 hairdressers, 51 pharmacies and 25 butchers. Cinemas rubbed shoulders with photographic studios. It was possible to down a Primus in one of the many cafés, while waiting for a local tailor to run up a suit. Markets in the camps were so well-stocked with vegetables, grown on tiny refugee plots, Zaireans sometimes headed out to the settlements to do their shopping. The refugees even ran their own transport service between the camps and Goma, using buses Japan had once donated to the Rwandan government. While hardly luxurious, life was certainly tolerable. With their vaccinations, regular diet and medical check-ups, the Rwandans enjoyed a higher standard of living than local Zairean peasants.

UNHCR and the myriad aid organisations who set up base in Goma ensured this was so. In the first days of the crisis, they had deluged the camps with food, plastic sheeting and utensils, not realising they were duplicating each other's work. The initial oversupply allowed community leaders to stockpile, providing them with the raw materials with which to jump-start the camp economy and trade with the locals. The sudden rush of funds did not stop with the stabilisation of the crisis. The aid agencies hired trucks and aircraft, rented local offices, warehouses and hotel rooms, took on translators, administrators and drivers. In the last nine months of 1994 alone, UNHCR and the aid organisations dedicated at least $336 million to the Zairean part of a vast refugee operation spanning the Great Lakes region, a sum that exceeded the Kinshasa government's total annual operating budget. Even if a share of that was spent outside the area on flights and logistics, what remained still constituted a heady injection of funds for a hitherto neglected provincial backwater.

For a president who had always used money to maintain his hold on the country, the financial influx into this 100-mile strip of land running along Lake Kivu marked a turning point. At a time when Gécamines and MIBA, Mobutu's traditional sources of ready cash, were barely operational, funds he could neither control nor appropriate came pouring into Zaire. For the army generals and Big

Vegetables who had once looked to Mobutu as sole provider, there were arms deals and security arrangements to be negotiated with the Hutu extremists, food and transport contracts to be struck with the aid organisations. Every transaction offered opportunities for bribes and commissions, sweeteners and the usual 'leakage', none of it granted at Mobutu's bequest. For a leader who depended on financial patronage for his survival, it was the final stage in a drawn-out process of economic marginalisation.

If Kivu's refugee camps taught Zaire's elite they no longer needed Mobutu to prosper, they also brought home to neighbouring states that he was no longer a leader they could do business with. Throughout his career Mobutu had played the game of befriending, sheltering or simply tolerating on his territory guerrilla groups dedicated to the overthrow of fellow central African leaders. The Hutu extremists determined to topple the Tutsi leadership in Kigali were to prove no exception. They had struck up solid friendships with Zairean army commanders who allowed them a free hand when it came to sabotaging the RPF's attempts to build a post-genocide society with a series of raids across the border. Bringing with them Rwanda's infectious ethnic hatred, they had also won Zairean backing for an operation to ethnically cleanse the Masisi region in north Kivu of local Tutsis, never popular with other Zairean tribes. The camps were feeling cramped, and the Hutus wanted a temporary homeland from which to prepare their planned invasion.

By late 1996, it was south Kivu's turn to be cleansed. The local deputy governor told the Tutsis from the Banyamulenge hills they were *persona non grata* in Zaire. For the Banyamulenge, who had seen their Tutsi brothers in Rwanda and Masisi slaughtered and driven out, it was tantamount to announcing a new genocide was about to be launched. It was not a development that took the new authorities in Kigali by surprise. They had watched the extremists establishing their fiefdoms in Kivu, had tried in vain to pre-empt the guerrilla raids from Zaire that left buses smouldering, schools machine-gunned, ethnic reconciliation a sour joke. Together with ally Uganda, they had complained repeatedly to Zaire, called on the UN to either move the camps away from the border or bring the hardliners to heel, hinted that they were considering unilateral action. But nothing had been done.

So, the Rwandans began infiltrating Tutsi fighters and weapons into east Zaire. In October 1996, at their instigation, four guerrilla movements announced the formation of the Alliance of Democratic Forces for the Liberation of Congo-Zaire (AFDL) in Lemera, south Kivu. In the fighting that followed, it took the rebel coalition and its neighbouring allies less than a month to achieve what the UN and Zaire had failed to do for two and a half years. As the interahamwe fled west, taking what followers they could, the extremists' hold on the camps was finally broken. UN plans for an international force to 'save' the Rwandan refugees trapped in the camps were quietly shelved. At every border crossing a multicoloured ribbon made up of refugees—bowed under their sleeping mats and cooking implements—stretched to the horizon. As the air once again filled with the sound of hundreds of thousands of feet brushing the earth, Rwanda's Hutus doggedly walked home.

CHAPTER TWELVE

The Inseparable Four

'Dictators ride to and fro upon tigers which they dare not dismount. And the tigers are getting hungry.'

—**Winston Churchill**

When the AFDL's representatives started calling the BBC offices in Nairobi in late 1996, claiming they would march all the way to Kinshasa, journalists dismissed them with a weary shrug as yet another unknown guerrilla movement, the length of its constituent acronyms only rivalled by its obscurity, making wild plans and farcical claims. Africa is full of them: they surface, splinter into factions—yet more acronyms—only to disappear with equal suddenness.

Anywhere else in the world, the AFDL story would have probably been one of raids on helpless villages, a few clashes with the army, limited annexations of land. A hotch-potch of credos, experiences and motivations, its membership ranged from communists to US-educated academics and village thugs. They had barely had time to work out either a clear structure or an ideological line when south Kivu's deputy governor pushed them into the limelight. Laurent Kabila, the spokesman-turned-leader, was a Maoist with keen commercial instincts, who had funded a fiefdom in eastern Zaire by smuggling out gold and ivory, a trade enlivened by an occasional spot of kidnapping of Westerners. Some of his colleagues thought they were fighting for the overthrow of capitalism, some for the survival of Zaire's Tutsi community, some for the end of Mobutu.

Sure enough, stories from Kivu soon began filtering through: of rape and looting, car-jackings and murder, hysterical fighters on the run. But on examination these turned out not to be atrocities committed by defeated rebels. At the first hint of an encounter with the

AFDL and their Rwandan and Ugandan allies, Zaire's hated army was grabbing what it could find, stealing the four-wheel drive vehicles owned by the aid agencies and heading for the interior.

As Kinshasa promised a 'devastating counterattack' that never materialised, town after town 'fell' to the AFDL, whose fighters, in their trademark black wellington boots, could barely keep up with the army's accelerating rout. Despite ineffectual UN appeals for a ceasefire, it became clear this was a war in which very little actual fighting was going on. Conquests began to follow a predictable routine. The incoming rebels would make it clear they intended to take a certain town. Alerted by the arrival of the first drunk army deserters that they were about to face a security crisis, local dignitaries would pool funds and lay on trucks or planes to evacuate the retreating FAZ. If the community could not afford the transport, it hit the road, more afraid of its own army's brutality than anything the rebels could do.

If Mobutu's regime could not quite believe what was happening, neither could the West. With the vantage point of historical hindsight, the telegrams sent by Daniel Simpson, US ambassador in Kinshasa at the time, show an extraordinary knack for getting it completely wrong. They are a measure of how thoroughly Zaire's diplomatic corps—like the country's population—had fallen under the Mobutu spell. 'The dramatic parts are almost over. The Rwandans have completed what they came to do,' Simpson told Washington in early November. Mid-month, with the AFDL still extending its campaign, he slapped down notions that the rebels enjoyed support outside the Kivu region. Any idea that the movement had supporters throughout the country was 'just silly'.

In early December, Simpson ruled out any risk of the rebels turning west and heading for the capital. 'A south Lebanon-type buffer state is all that Rwanda and Uganda have signed on for,' he said. In January 1997, during a lull that preceded Angola's army joining the anti-Mobutu onslaught, Simpson concluded: 'The Rwandan-Uganda backed rebellion in the east of Zaire is falling apart.'

Whatever Big Vegetables in Kinshasa came to believe, the AFDL's lightning advance was not the result of massive logistical support from Anglophone Western nations determined to destroy their former ally. Zaire's security system was collapsing like a maggot-eaten fruit. As village after village greeted the AFDL 'liberators', the campaign Rwanda and Uganda had launched to eliminate a border problem transformed itself into something else entirely: the takeover of a vast country. To misquote Churchill, never in the field of military history had so much territory been captured by so few with such little effort.

Zaire's national embarrassment of an army traced its roots back to the Force Publique of the colonial era, which, while hated, had been ruthlessly effective. Its officers had been taught in the best military academies of the US, France, Belgium and Israel, trained by experts from Germany, Egypt, China and South Korea, and supplied with some of the most sophisticated equipment ever seen in Africa. France had provided a batch of Mirage jets; the CIA technicians to maintain its aircraft. In the 1970s it had been regarded as credible enough to contribute to international peace-keeping operations. Other African nations had even sent their officers to Zaire's centres of military excellence for training.

During the first decade of his rule the army had been Mobutu's pride and joy—modernised, expanded and restructured. The former sergeant's original ascension was premised on his success in curbing an army mutiny, his understanding of what made the ordinary subaltern tick. His ability to stay in power long after support had waned had depended in part on public dread of the men who were now quietly stripping off their uniforms and melting into the crowd. No one, surely, could be more aware of the importance of army morale than Mobutu. So what had gone wrong?

I met the man who thought he knew the answer in the bar of the Intercontinental Hotel in Paris. Outside, the sun was blazing down on tourists scrunching the gravel of the Tuileries Gardens, but here it was dark. As a result, his photosensitive lenses had turned clear and I could see his eyes, usually hidden behind dark glasses. He was slightly smaller than I remembered and for a moment I wondered how this quiet, soberly dressed man—his only visible extravagance a diamond-studded gold watch—could ever have become a figure of such controversy.

But it was when he started talking that I was reminded of his nickname. It is not something one likes to mention in his company, but Honoré Ngbanda Nzambo Ko Atumba is commonly known amongst Zaireans as the 'Terminator', a reference to the horrors carried out by the 'owls', the sinister force responsible for night-time interrogations and disappearances which cracked down on opposition activists and troublesome students during the five years he spent as head of the intelligence service. And there is something about the Terminator's voice that strikes a chill to the heart. It is clipped, slightly nasal and instantly recognisable. His French is impeccable, his phrases wind their way through subsections, qualifications and subtleties, pointing to a coldly precise brain behind. It is a sophistication which has determined the course of his life.

He was a brilliant, seminary-educated young man when he was talent-spotted by Mobutu. Presenting the student body's complaints to the president, he made such a good case Mobutu told the head of his intelligence services to follow his academic career and recruit him on graduation. At his mother's suggestion, he let drop ambitions of becoming a priest, while holding on to the Christian faith with peculiar fervour. The secular world called. It was to bring him decidedly unmonastic levels of wealth while sharing some of the characteristics of the priesthood: a familiarity with occult forces and intimate secrets, an awareness of the machinations unseen by the common man, and, finally, privileged access to a supreme being held in awe by mere mortals. As the Jesuits proved during the Inquisition, spirituality can go hand-in-hand with ruthless single-mindedness when the individual is convinced his cause is just.

Several foreign assignments were followed by a posting as ambassador, the directorship of the SNIP intelligence service, three stints as Defence Minister and nomination as Mobutu's special security adviser. The Terminator succeeded where Janssen, the white playboy venturing out of his depth, had failed. He was privy to the president's most secret thoughts, entrusted with the most delicate of diplomatic missions. His role, which won him the sobriquet of 'Special' from Mobutu, made him a natural target of Kinshasa's scurrilous rag-sheets. Journalists speculated about his business interests, cartoonists depicted him—with sideburns and signature sunglasses—as a kind of thuggish spiv hatching dark plots with fellow aide Vundwawe Te Pemako, the two real powers behind the throne left vacant while Mobutu disported himself in Gbadolite.

It is that impression Ngbanda had set out to dismantle with his account of Mobutu's last moments, written from comfortable exile in South Africa. While talking of his beloved 'Marshal' with intimacy and affection, the Terminator nonetheless delivers a series of killer punches. Painting a pitiful picture of a vacillating president, surprisingly naïve and often in floods of tears, Ngbanda's message is clear. Despite his key position as presidential confidant, he would never do more than recommend. His advice was often ignored or applied only after the moment had passed by a head of state overtaken by events. The ensuing débâcle could be blamed on the family, the generals, the West, but not, repeat not, on the president's security adviser.

It is a stance that infuriates many members of the former elite, including Mobutu's own family. 'It is just too easy for the former aides to keep saying: "We gave Mobutu good advice but he never followed it," fulminated son Nzanga, proud of the fact that he has had no contact with the Terminator since leaving Kinshasa. 'If a president doesn't listen to your advice for ten years, you should resign. I think a bit of *mea culpa* would have been appropriate from people like Ngbanda.'

And there are moments in Ngbanda's narrative when the rewriting of history to ensure he emerges unscathed becomes a little too blatant. Given how close he was to the pulsating heart of power, it is surprising how often the Terminator is taken by surprise, how frequently foreign ambassadors or heads of state have to spell out to this insider facts all Kinshasa has already suspected. Yet put to one side all the carefully paraded innocence, all the self-justification, and the powers of analysis that so impressed Mobutu make themselves felt. The Terminator's critique of the Zairean armed forces—that body that turned on its own society and tore at its own entrails like some rabid animal—is too well-argued for even his worst enemies to do much more than nod in glum agreement.

For Honoré Ngbanda, the problem could be traced back to the management technique on which Mobutu had founded his regime. Pursued through the decades, the tactic of divide and rule emerged as little more than inaction turned into an art form, a vacuum where decision-making should be. Yes, it offered stability of a sort, but this was the stability of a taut elastic, the calm at the heart of a hundred forces tugging in different directions. To really achieve something, to build a bridge, pave a road or win a war, such forces must, however briefly, pull in the same direction. Just as he made concerted action impossible at a political level, Mobutu, the two-time *coup*-maker, was careful to ensure the armed forces never boasted a unified command structure that could be exploited by a popular rival.

His attitude to the army underwent a fundamental change in the late 1970s, when he woke to the danger represented by a disciplined, motivated force. In 1975 a group of officers from the central Tetela region were arrested on charges of plotting a takeover. Three years later another alleged *coup* attempt was foiled. Thirteen people were executed and more than 200 officers from Kasai, Bandundu and Shaba purged.

Mobutu had already been pushing into retirement older officers who had helped him seize power. Now the army lost a huge swathe of its brightest and best-trained. Kasaians were regarded as untrustworthy, hailing as they did from the province of Tshisekedi. But Bandundu and Shaba were also declared off-limits in recruitment drives as the armed forces acquired an increasingly equatorial tinge.

The tribalisation of the armed forces was not new. Like all colonial masters, Belgium had tended to classify Congolese ethnic groups into 'war-like' and 'non-war-like' categories. Mobutu's tribesmen had been labelled natural warriors and, as a result, already held a disproportionate number of army posts. Mobutu now took that principle to new extremes as he ensured the security forces' top echelons were ethnically predisposed to his rule.

The West kept pouring funding, equipment and experts into Zaire in an attempt to establish a respectable army. But it served little purpose. Increasingly, experience and professionalism were regarded as irrelevant when it came to doling out top jobs, allotted to people from northern Haut Zaire or Equateur province. Soon, even that limited recruitment pool narrowed to the North Ubangi region from which the Ngbandi hailed. With the Special Presidential Division (DSP), recruited overwhelmingly from the Ngbandi, the principle was taken to its logical extreme. Outsiders in Kinshasa, regarded with fear by the local population, their loyalty was virtually guaranteed. While publicly preaching Zairean nationhood, Mobutu only trusted his own tribe, it was clear, with his safety.

Amongst the Ngbandi, members of Mobutu's family did best, with general's stars doled out generously to cousins and brothers-in-law. But Mobutu knew his own relatives too well to feel entirely at home even with that arrangement. To distract the generals, he kept them uncertain of their positions, constantly bickering amongst themselves. Using a method perfected by Adolf Hitler, Mobutu would give similar responsibilities to bitter opponents, then sit back and watch the sparks fly. 'Each defence minister or general had, at the head of the army or in Mobutu's entourage, an "opponent" against whom he had to defend himself: Bumba was attacked by Molongya, Singa was assailed by Lomponda; Likulia insulted Eluki; Mahele put Eluki through the hoops while Singa, back at Defence, was targeted by those who had nominated Likulia as his secretary of state . . . and so it went on,' recalled the Terminator. Such rivalry, he stressed, was not a regrettable accident, it was the very basis on which the armed forces were run.

As the generals jousted, myriad elite forces sprang up, each answering directly to Mobutu. Every general sought to shore up his position by recruiting as many young men from his own village as possible and pressing for repeated upgrades. Promotion came at a stratospheric pace. By 1997 the armed forces had become ludicrously top-heavy, boasting fifty generals and over 600 colonels.

Riding the tiger, Mobutu's role was more that of a Mafia 'capo dei capi', focal point of several highly tribalised gangs, than supreme commander of the armed forces. He allowed one elite to be built up, then, when it seemed in danger of posing a real challenge, switched resources and patronage to another. Hence the multiplication of special units and security organisations, often vying for identical duties: the DSP, the Garde Civile, SARM, the Kamanyola division, the para-commandos, the 21st brigade, the 31st brigade, SNIP, and, bringing up the rear, the gendarmerie, police and regular Forces Armées Zairoises (FAZ). Despite the sheer size of the country, most of these elites were kept close to Kinshasa, rather than patrolling the borders. Their positioning reflected their role. The Zairean army was not aimed at resisting external attack. It was an internal security machine whose sole *raison d'être* was protecting the president.

If the elites at least enjoyed high pay, decent equipment and the social respect born of fear, the regular army was treated like dirt. The Ngbandi generals could never muster much military experience between them, but they knew how to make money. And the simplest method was to appropriate the contents of the trucks which arrived with the troops' salaries each month. The practice explained the curious fact that no one ever knew how many men in uniform Zaire actually boasted. The generals demanded pay for 140,000 men, almost double the 80,000 estimate of most experts. The government knew it was being cheated. But when Ngbanda, as newly appointed Defence Minister, tried to organise a head count to end the double-billing, he discovered what formidable opponents the generals could be. The generals told their troops the new minister had suspended their pay for indiscipline, then warned Mobutu a mutiny was about to explode. Mobutu begged Ngbanda to abandon the idea.

With weapons but derisory levels of pay, the soldiers behaved as could only be expected. They emerged from their barracks to prey on their own citizens, building on a tradition firmly established by the Force Publique. Encouraged by its president to 'live off the land', the FAZ gradually disintegrated into a force adept at hijacking cars and stealing beer but utterly unskilled in the business of war. For those who have not lived in a country fallen victim to a rogue army, the extent to which the phenomenon transforms a society is impossible to imagine. The heart of a white, middle-class Westerner does not automatically miss a beat at the sight of a military uniform. After my time in Kinshasa, mine did. I had made the necessary mental leap, from viewing an army as a society's shield to regarding it as a testosterone-charged time-bomb, primed to blow apart its own community.

These angry young men in their pimps' sunglasses, Kalashnikov cartridge clips Scotch-taped together, trousers held up with bootlace, infiltrated every aspect of life—Article 15 at its ugliest. At 'roadblocks' consisting of a frayed piece of string stretched across the tarmac they lounged drunk, levying 'taxes' on traders taking goods to market. In bars they ordered customers to buy them beers, at taxi stops they clambered fully armed into cabs, forcing 'protection' on frightened passengers.

Each month, tension would rise as the troops' paltry salaries were exhausted and the scrounging became more blatant. Then came an uncertain few weeks of rumours. So-and-so knew for certain the army had been paid. But so-and-so said the troops were unhappy over the amount. All it took was a power cut, and panicking businessmen would be on the Telecel, warning that another round of pillaging had begun.

Like a surly adolescent who bullies his own parents, the army held Zaire hostage. And the Zaireans, so proud of their tradition of non-violence, so steeped in passivity, tolerated it. 'The Zaireans must take a large part of the blame,' said a doctor who worked sixteen years in Kinshasa. 'If just a few of those soldiers swaggering around the Cité had had their throats cut in the night, it would have made a difference. Instead, the Zaireans let the soldiers live amongst them.'

But Mobutu also eventually paid a price for such sabotage. He was like a poker player with a worthless hand, hoping no challenger would be gutsy enough to call his bluff. His own courage was never in doubt. An expatriate who accompanied him to several war zones remembers him standing broad-shouldered as the bullets whizzed around, shaming quaking soldiers into action. But in the list of the FAZ's military engagements, victories take some finding. So rare was the event, in fact, that when it did occur, it was commemorated with nauseating frequency. The presidential yacht, an army division and Kinshasa's sports stadium were all named after the eastern town of Kamanyola, where in 1964 Mobutu and his men captured a rebel-held bridge.

He rendered the FAZ so incompetent, he had to rely on outsiders to do his real fighting. Moise Tshombe, who recruited 'les affreux' (the terrible ones) to back up his post-independence Katangan secession attempt, set a precedent Mobutu was happy to follow. When the going got tough, US, French, Belgian, Cuban, South African and Rhodesian mercenaries got going. Know-how was not the only thing Mobutu was after when he signed up the likes of Colonel Bob Denard and Jean Schramme. He was also hiring a myth, a concept of ruthlessness, because he believed the colonial experience had left most African troops imbued with a colossal inferiority complex, convinced a white man with a gun would always be the equivalent of twenty home-grown fighters.

But renting mercenaries was only necessary when Mobutu's foreign friends could not be counted on to win his wars for him. And most of the time they obliged. In 1977, when just 1,500 Katangan rebels routed the FAZ in the strategically key south, France flew in Moroccan troops to win the First Shaba War. When a similar attempt was made a year later, it was snuffed out by French foreign legionnaires and Belgian paratroopers, followed up by a pan-African peacekeeping force. And when the army itself seemed about to unseat Mobutu in 1991 and 1993, French and Belgian troops once more helped save the regime, with the former patrolling the streets of

Kinshasa while the latter lined up along the Brazzaville frontier, sending a message anyone planning to seize the opportunity to topple Mobutu could not misunderstand.

But the two 'pillages' were a sign that the tactic of divide and rule had run its course. The anarchy Mobutu had nurtured in self-protection had reached a point where it risked bringing the whole regime crashing down. It was a message, Ngbanda claimed, Mobutu decided not to hear. Instead of reigning in the generals, he doled out promotions. Rather than discipline the mutinous troops, he granted salary increases—a fairly pointless exercise given that few ever saw their full pay packets.

His nemesis was to take the shape of a small clique of men sporting generals' stripes that owed more to links of marriage, friendship and family with the president than professional experience. Popularly referred to as the Inseparable Four, they were in fact a group in which two generals, Nzimbi Ngbale, Mobutu's cousin and head of the DSP, and Baramoto Kpama Kata, commander of the Garde Civile, were the brightest stars, with General Eluki Monga and Admiral Mavua Mudima as smaller satellites. Once Mobutu was conveniently absent in Gbadolite, the Inseparable Four swiftly emerged as the real powerbrokers in Kinshasa. 'They went everywhere together, from official appearances to private gatherings,' commented one general, Ilunga Shamanga, who as a Kasaian remained outside the magic circle. 'What seemed a wonderful example of solidarity and cohesion was in reality nothing more than a criminal association.'

In a slip that particularly irked Ngbanda, Mobutu had allowed the generals to wrest control of the intelligence services in 1990. It was an error with enormous long-term consequences, because it meant they could feed the president with misleading data about conditions on the borders and troop morale. From then on, the neutral information the president needed to take sober decisions was tainted. While opposition newspapers obsessed about Mobutu's motives and even foreign diplomats seemed entranced by the myth of presidential power, the story had already moved on. Rolling around town in jeeps with tinted windows, the generals had their hand in every financial scam, from diamond dealing to the importation of forged zaire notes. They were even pushing for direct political involvement.

For the Terminator, the moment when he realised it was the Inseparable Four, and not his boss, who now called the shots, came when the generals took umbrage at not being consulted over who should head the central bank and state enterprises, potential sources of illicit income. They sent troops and tanks to surround each building, preventing the new chief executives from reaching their offices. Fuming, Mobutu summoned the generals to his residence. 'Either you free up those offices or I resign,' he shouted. They obliged, but the way Mobutu had delivered his ultimatum shocked his entourage into stunned silence. 'He had not threatened to sack the generals or discipline them for insubordination. Instead he was the one who had threatened to resign,' recalled Ngbanda. 'I understood something had changed in his relations with the generals: the balance of power. I had the profound conviction that the death knell had sounded for Marshal Mobutu's regime.'

More significant in the grand scheme of things than the generals' thwarted political ambitions were their commercial interests, particularly the chutzpah they showed in selling off the contents of the national armoury. General Ilunga recorded the near-comic moment in September 1995 when he learned that Zaire's fleet of Mirage fighter jets, nominally sent to France for maintenance, had been quietly sold a year earlier. When Mobutu asked him to investigate, he was told the Mirages had been surrendered to allow the president's helicopter fleet to be modernised. The new helicopters never made an appearance.

But usually, the trade was less ambitious: ammunition and rifles, sold to the guerrilla movements who had established their bases on Zaire's barely policed frontiers, irrespective of their friendliness or hostility to the Mobutu regime. There is something of the inspired insanity of *Catch-22*'s Milo Minderbinder—the mess officer so obsessed with a bargain he arranges for American bombers to flatten their own air base on the Germans' behalf—about the way in which commanders in Kivu, despite clear signs a conflict was looming, happily sold arms to the very AFDL insurgents who would eventually chase them from the area, then set fire to storage warehouses to conceal the hole in supplies. Showing all the far-sightedness of a man handing a neighbouring arsonist a canister of petrol and some matches, the generals could not resist clinching the shady deal, even when it meant jeopardising their own futures.

As the AFDL began crossing the country in what was to prove one of the swiftest campaigns in modern African history, the generals called for defence budgets to be upped, then siphoned off the best of deliveries, leaving the FAZ with ammunition that did not match its rifles, second-hand equipment from Eastern Europe long past its prime. Maybe the generals had begun to believe their own reassuring report to Mobutu. Maybe they were too stupid to think through the consequences of their actions. 'To us that kind of behaviour seems incomprehensible,' marvelled an ambassador. 'They were sabotaging their own campaign. But you have to regard these people as gangsters rather than politicians. And a gangster tries to make money until the very last moment.'

The trade was not only damaging because it emasculated the FAZ. Mobutu's long-standing support for such guerrilla groups—particularly his close friendship with Angolan rebel chief Jonas Savimbi—had been a sore topic with neighbouring nations for decades. Affected countries assumed that either this arms trade was taking place with Mobutu's backing, or that the generals were acting on their own behalf, a sign he no longer controlled the situation. Either way, it was time for Mobutu to go.

Lambert Mende, Transport Minister in Mobutu's last administration, logged how seven companies owned by the generals and members of Mobutu's entourage were still flouting a government ban on arms flights into UNITA-held territory as the AFDL rebellion gathered ground. 'The Angolans said "if this continues we will join the war". But they continued, and the Angolans joined the war.'

So it was that first Rwanda, then Uganda and finally Angola were eventually to join forces with the AFDL in a momentary coalition of regional interests never before witnessed in Africa. Zaire became the terrain on which alien forces worked out ancient grudges, with the locals swept along for the ride. Tutsi troops from Rwanda hunted down the interahamwe in the equatorial forests, killing untold numbers of Hutu refugees in the process. Angolan soldiers seized an opportunity to track down the UNITA fighters who used Zaire as a rear base. Zambia co-operated by letting the AFDL cross its land to win access to the south; Zimbabwe and Eritrea supplied arms and Tanzania turned a blind eye to the rebel training camps on its territory.

When the rebel campaign first began, Zaireans waited for Mobutu to send the elite forces they had heard so much about to the east. No one, after all, could expect the FAZ to stand up to protagonists fuelled by the loathing Rwanda's ethnic divide seemed to breed. As a waiter in Kivu once confessed to me, in one of those endearing Zairean moments of self-insight: 'We Zaireans may be thieves. But those guys over there,' he said, jerking his head towards the border, 'those guys are killers.'

They waited and waited. Was Mobutu playing some kind of clever tactical game? Was he saving the DSP for later? The answer was much simpler, according to the Terminator: despite repeated orders from Mobutu, not one of the Inseparable Four ever agreed to follow the example set by the youthful Mobutu and go to the eastern battlefront. They had no intention of risking either their own lives or the forces they regarded as priceless tools of financial extortion.

In desperation, Mobutu applied the methods that had saved him in the past. He appealed to his Western friends. France, always the most loyal, pushed for the UN to send an international force to east Zaire to 'save' the Rwandan refugees. But the project was cancelled when most of the refugees flooded home and in the wake of the outcry over Paris's support for Rwanda's toppled genocidal regime, a solo operation was out of the question. Both Belgium and the US publicly washed their hands of him.

Cut off in Gbadolite, Mobutu had failed to appreciate how dramatically *realpolitik* had altered in a post-Cold War world in which leaders recited the mantra of human rights and democracy. 'This error in judgement was one of Mobutu's most fatal. God knows how often we had discussed the issue with him,' said the exasperated Ngbanda. 'But all attempts to explain the objective basis for the change in US politics towards him were violently rejected.'

Belatedly registering that the Inseparable Four were more hindrance than help, Mobutu sacked Baramoto as chief of staff in December and replaced him with Donat Lieko Mahele, an Equateur general, like them, but one with a level of professional competence absent in his colleagues. But even Mahele could not undo in a few months the sabotage of decades. Although given full powers by Mobutu, Mahele soon found that Nzimbi and Baramoto were telling their special units to disobey his orders. At gunpoint, they refused to hand over vital weapon supplies. The regular army was still not being paid. As ever under the Mobutu system, it was impossible to say what came from the president and what was the work of subordinates, particularly as Mobutu was abstracted and hesitant, going back on key decisions, responding achingly slowly to events. 'I know the president,' a frantic Mahele railed at the Terminator, only days after his appointment. 'He is beginning to play the game of turning one man against another. I want no more of it. If he doesn't give me the weapons I need, I will resign.'

In a resort to tradition, the white mercenaries were summoned, in an operation co-ordinated by the French secret service. But as Machiavelli could have told Mobutu, mercenaries—unlikely to take enormous risks, their loyalty always open to question—are a far from ideal solution. 'The thing to bear in mind about mercenaries is that so many live to write their memoirs,' was one military analyst's sardonic comment.

Its ranks swelled by Serb psychopaths fresh from Bosnia's killing fields—men who were to be arrested in Belgrade nearly two years later, accused of plotting the assassination of Yugoslav President Slobodan Miloševic—the mercenary force had trouble liaising with the FAZ. There were language problems and several 'friendly fire' incidents. It was never clear who the mercenaries answered to: Mobutu, the prime minister, Mahele or the French ambassador. While succeeding in terrifying the local population with their atrocities, the mercenaries signally failed to pump new life into the FAZ campaign by the time they fled.

In a last gasp, Mobutu begged his African allies for support. Nigerian military leader Sani Abacha, mindful of the role Mobutu's troops had played fighting the Biafran secession, was amongst the few to respond positively. Yet nothing ever came of the plan. When it was far, far too late, Ngbanda discovered Abacha had been told to stand down by a Francophone African leader purporting to speak on Mobutu's behalf. The region itself had decided the era of the dinosaur was over.

A number of myths were exploded during the seven brief months that separated the birth of the AFDL from the storming of Kinshasa. Already battered, the image of the invincible white mercenary in Africa finally collapsed. The notion that France would always send in forces to shore up its African friends, however corrupt, was overturned. Above all, the belief that Mobutu held the key to the armed forces, a key that would keep him in power no matter how unpopular he became, evaporated.

By March 1997 the AFDL had taken Kisangani, a military turning point. The following month came Mbuji Mayi, then Lubumbashi, where Kabila received a warm welcome from his fellow Katangese. As FAZ pulled back, leaving the interahamwe and UNITA rebels to do the real fighting, Kinshasa was being cut off from the country's mineral resources, a capital with no hinterland.

Maybe it was as the first boats laden with army deserters began landing in the capital that the generals stopped believing their own lies. As checkpoints were set up on the highways into the capital—manned by soldiers far more suspicious of the army unit down the road than any rebel infiltrator—the men Mobutu hoped would never betray him because of friendship, ethnic loyalty and family ties began to plot against their former champion.

Had this succession of reversals and betrayals occurred a year earlier, when he was still in good health, Mobutu might have been able to call upon his vast reserves of cunning and pull off one last diplomatic *coup*. His political demise had been announced time and time again, only for him to emerge phoenix-like from each crisis. But Mobutu was a sick man. Scenting the sweet odour of decay, his enemies crowded around like hyenas snapping at a wounded leopard.

CHAPTER THIRTEEN

Nappies on the floor

'The first opinion which one forms of a prince, and of his understanding, is by observing the men he has around him, and when they are capable and faithful he may always be considered wise, because he has known how to recognise the capable and to keep them faithful. But when they are otherwise one cannot form a good opinion of him, for the prime error which he made was in choosing them.'

The Prince
—Niccolò Machiavelli

By 1996, the young athlete who could outsprint his contemporaries and revelled in such physical feats as parachuting had undergone something of a transformation. Mobutu was sixty-six and three decades in office had taken their toll. His hair was still dark, but the colour now came from a bottle, applied by a Lebanese hairdresser flown to Gbadolite for the task. His face seemed to have registered every sleazy deal, every moral compromise along the way. The eyes were hooded and his features had coarsened, the pouting lips settling naturally into a downward droop that hinted at scepticism and disappointment.

Mobutu did not look well and the CIA, in one of its classic pieces of misinformation, confidently informed Washington that the president was suffering from AIDS. Given his reputation for exercising a presidential *droit de seigneur*, the diagnosis must have seemed plausible. He was indeed ill, but he was suffering from something far more prosaic but just as deadly in the long term: prostate cancer, one of the most common causes of death in elderly men.

If the cancer is caught early and properly treated, as it was with French president François Mitterrand and Archbishop Desmond Tutu, the patient can last for years. But diagnosis is not easy. The physical examination, which involves inserting a finger in the rectum, is not of a kind doctors lightly press on a grouchy African autocrat. By the time it is clear the patient has cancer, the disease may have already infiltrated surrounding tissues, triggering secondaries in the spine and lungs.

The first family and aides knew of the problem was when Mobutu was whipped into surgery in a Swiss clinic in Lausanne in August 1996 after a routine medical check-up revealed an abnormality. Unless Mobutu had managed to keep such a momentous secret entirely to himself for several years, which seems unlikely, we must assume, therefore, that he learned of his illness almost at the same time as the outside world: in other words, just when he was confronting the biggest threat of his career. The immediate decision to operate and the rapid course the disease then followed make it clear this was no Mitterrand-style slow-moving cancer. It was spreading across his body as quickly as the AFDL were gobbling up his territory.

The medical treatment he underwent must have been a reminder that the great Guide, the all-seeing Helmsman, was a mere mortal, after all. Because it involves surgically removing the prostate or hormone treatment to reduce the level of testosterone in the blood—effectively chemical castration—impotence is the norm. Patients can also experience 'feminisation', losing facial hair and developing breasts. For an African symbol of virility such as Mobutu, the man who could 'cover all the chickens', this would have been hard to bear. Even worse humiliations loomed: because of the prostate's position, it is hard to remove it without damaging the bladder or urethra, so surgery can cause incontinence.

It was not surprising, given the draining fatigue produced by radiotherapy, that Mobutu dithered and dallied more than ever before, unable—to the bafflement of those around him—to make the split-second decisions on which his survival now depended. No wonder, given the battering his self-image had sustained and the sudden prospect of an early death, that he seemed at times paralysed by depression. Ngbanda, the special adviser, increasingly had the sense he was dealing with two different individuals: one who showed all the alertness and dynamism of the old Mobutu, another morose, sunk into lethargy.

Yet the president made a superhuman effort to put his own physical problems to one side. In late December he staged a symbolic return to Kinshasa, the capital he had shunned for so many years. Those who watched the motorcade said it was like the 1970s all over again, as thousands of Kinshasa residents, frightened of what this 'foreign invasion' in the east spelled and curious to see how sick 'Papa' really was, gathered to cheer the leader who had come to share their ordeal. Invigorated by the unexpected adulation, Mobutu stood upright in his open-roofed limousine for the length of the 35-kilometre trip from the airport to Camp Tsha Tshi, holding his presidential cane aloft and basking in the cheers: quite a feat, in the African heat, for a healthy mortal, let alone a cancer-stricken sixty-six-year-old. He was left so hoarse from hailing the crowd, he could barely read his speech later that evening, in which, in a voice cracked with emotion, he promised to meet the population's expectations.

It was to be his last such moment of glory. Public support waned, as Mobutu failed to produce a miracle solution, and so did the president's energy. When he returned to Kinshasa from Nice in late March 1997, dragging himself away from the doctors and nurses, there were no applauding hordes. Instead the event was overshadowed by a strange incident at the airport. The presidential plane had landed, the red carpet had been rolled out and members of the government were lined up on the tarmac. They waited and waited, but Mobutu did not emerge. Eventually, the press was shooed away and the cabinet told to disperse, prompting immediate rumours that Mobutu had either died on the flight or was petulantly refusing to meet Kengo Wa Dondo, the prime minister he was known to detest.

In fact, Mobutu's limbs had seized up during the long flight and he was physically incapable of standing upright until massaged by his carers. When he finally descended from the aircraft, away from the media's unforgiving eye, he was leaning heavily on his wife and the car had to be brought to the foot of the stairs. His brief respite was over, the cancer was beginning to bite.

This was a turning point in the career of a man who had lived in the public spotlight since his thirties, dividing his time between army generals, cabinet ministers and foreign VIPs—presenting an image of iron invincibility. Few African presidents had been so thoroughly filmed and photographed. 'With Mobutu, there were always cameras around,' a World Bank man once told me. But as we approach death, we all shed the inessentials and decide who and what really matters. For Mobutu that meant completing the personal voyage he had started in Gbadolite: he turned inwards, searching for the warmth and support only blood ties could provide. As the bleak realities of life crowded in, he wrapped himself in the soft cocoon of his family, becoming once again a very private man.

For one young man, yearning for intimacy with the legendary figure who was his father, this was the moment he had been waiting for. Struggling to come to terms with a sinister patronym, Nzanga Mobutu knew this was the last chance for connection with Mobutu Sese Seko.

I couldn't control a slight tremor of excitement when voicing the name. It was, it seemed to me, a bit like asking whether Mr Genghis Khan had called or Mrs Caligula had left any messages. But the clerk at reception, a man with the suave air of one who had witnessed comings and goings normal mortals could only dream of, never blinked. He scanned his huge bookings diary with the casual confidence of a pilot assessing his control panel.

'Mr Mobutu? He's not staying with us, is he?'

No, I said, but he was expected. In fact, I understood he had hired a special room so we could chat in peace.

'Ah yes, he hasn't arrived yet. But if you wait over there, I'll let him know where you are when he arrives.'

He entered the lobby of the discreet four-star hotel off Paris's Place Concorde, a place he said was a favourite because of the 'English feel' lent by the leather-bound volumes on the shelves, dressed in a slate-blue suit whose elegance was so understated it had clearly cost a great deal of money. He was sniffing delicately, nursing a cold and blinking back sleep. Insomnia, he explained slightly sheepishly, was something he shared with his late father. Like Mobutu, he tended to suffer from restlessness in the night, then waves of drowsiness in the day. 'I should really take siestas. But it's not convenient.'

Insomnia wasn't the only thing Nzanga Mobutu, Mobutu's son by his second wife Bobi Ladawa, had inherited from the late head of state. At twenty-nine, Nzanga had the fleshy good looks, the pouting lips of his father, with his mother to thank for his light skin and doe eyes. 'My brother and I and my son, we all look exactly like my father. He must have had strong blood.' It is an inheritance, he said, he had no intention of turning his back on, whatever awkwardness it may bring. 'I'm very proud of the name that I bear,' he insisted. Indeed, Nzanga had treasured every moment spent alongside his father at the end of his life, however dangerous it had proved. Like a dry sponge plunged into water, he soaked up the trust and affection of the man he venerated, revelling in this father-and-son communion before death snatched the opportunity away for ever. 'Those last two years were so important to me. Every word he said, every thing he did, it was like a lifetime packed into a moment. Being by his side was the best university a man could ask for. Because before that we had only seen him on vacations. And he wasn't a man for vacations.'

Being a president's son can never be easy, but in the case of a man possessed of huge wealth and no shortage of enemies, it was doubly difficult. Mobutu always revelled in the company of children, and he would have loved to have lived surrounded by a sprawling extended family. He also wanted his offspring to retain links with home, a feel for African village life, rather than joining the rootless club of glitterati washing around Switzerland's ski chalets and Riviera resorts.

But against all this had to be weighed the constant threat of kidnapping and the embarrassing fact that Zaire could provide neither the education nor the health treatment he wanted for his children. The solution Mobutu picked was to farm his offspring out to former colonials, Belgian couples trusted for their complete discretion.

Under assumed names, Mobutu's children attended school and university in Europe and the US, returning to Zaire when term time allowed.

Nzanga left Zaire at the age of six and spent ten years living with a Belgian colonel who was a stickler for punctuality. 'He was very tough. He'd say "before time is too early and after time is too late". At the time I didn't appreciate it, but now I realise it was good for my education.' At Easter Nzanga went on school exchanges to Britain, visits that gave him a smattering of the English he now speaks fluently. Other holidays were spent in Gbadolite, but even then his father, travelling incessantly during the era when he ranked as an important non-aligned statesman, friend of both the capitalist United States and communist Romania, was often away. 'We relied on my mother. She played the role of father and mother at the same time,' said Nzanga. 'We missed him terribly. We really lacked a paternal presence. For my father it was work, work, work, all the time. Even when we were at the table he would be receiving visitors and holding meetings. He had no personal life. Which is why I want to be around my own two children a lot while they are growing up.'

On the surface, Nzanga's life was now one of gilded ease. Travelling freely between Brussels, Paris and the family base in Rabat, he had no problems with the European authorities as long as he made clear he was not planning to demand political asylum. Building on a Canadian degree in communications, he set up a communications group although, as one of the heirs to the Mobutu estate, it seems unlikely he will ever actually need to work.

When he talked about the president his voice thickened with pain, and it was clear that the events of 1997 had left their mark. He was obviously still mourning the father he belatedly came to know. *De facto* guardian of the Mobutu flame, he knew the risks he ran in talking to a journalist. But if there was any chance to contribute to a more nuanced, a more generous picture of the man whose crimes had been denounced across the world, he wanted to seize it.

When historians came to re-examine Zaire's ills, Nzanga was convinced, they would absolve Mobutu of much of the blame and focus instead on those he nicknamed 'the bloodsuckers' as a child: the aides and army men, premiers and ministers who manipulated the president, only to portray Mobutu as sole, misguided decision-maker when events turned sour. 'I call him the tree that hid the forest. I used to tell him to be careful of those around him. But he always thought of himself as a tribal chief. He would say, "I must cover this", and take responsibility for actions of others. He took everything on his own shoulders. He did not believe in passing the buck.'

The last battle in the long 'war of influences' that raged around Mobutu for three decades was staged in surprisingly simple surroundings for a head of state associated with often laughable levels of personal luxury. Mobutu's last months in Zaire were spent in a modest grey villa in the cool of the hills overlooking the first cascade of Stanley Falls, with a clear view across the river to the glinting skyscrapers of Congo-Brazzaville. Here he ensconced himself with Nzanga as his spokesman, twin sisters Bobi and Kossia as emotional supports, his son Kongulu—the feared 'Saddam Hussein'—looking after security, daughter Ngawali, the infamous Uncle Fangbi and his personal doctors. To their fury, the advisers whose services Mobutu had once called upon were now rendered virtually redundant, as increasingly he left family members to man the presidential Telecels. 'He no longer listens to us,' the head of the MPR party despaired. 'He only listens to his family, and they are pursuing their own agendas.'

There had been a time when Mobutu had groomed one of his sons—Niwa, widely regarded as one of the smartest of an original brood of seventeen children—for a role in politics. But with Niwa's death of AIDS any attempt to found a ruling dynasty was abandoned and Mobutu sought instead to keep his offspring out of the world that had disillusioned him. For Nzanga—'a garter snake in a nest of cobras', in the words of one ambassador—this was to be his first venture into Zaire's intrigue-laden political arena.

The villa in Kinshasa had pale echoes of Gbadolite magnificence: peacocks paraded the lawns, monkeys scrabbled at their cages and fountains played in ornamental gardens where lizards in lurid shades of purple and orange basked in the sun. But the reality which Mobutu had managed to keep at a distance in the forest lay just outside the grey railings surrounding the house.

The first presidential residence of independent Congo, the villa lies smack in the middle of Camp Tsha Tshi, main barracks of the DSP, a stone's throw from the washing lines hung out by the officers' wives. Only in this enclave within an enclave, surrounded by his clan's warriors, with a helicopter on standby to lift him to safety, could the besieged president now feel safe.

It was to this residence of convenience, not a place that Mobutu himself had ever called home, that the negotiators came, delegation after delegation, in their Mercedes and wailing motorcades, trying—as tense days ran into even tenser weeks—to answer the question which was baffling Kinshasa's inhabitants, Western governments and Mobutu's African allies. With the rebels marching unstoppably towards the city, with the Big Vegetables and their families heading out of town, with even Ethiopian Airlines, that most unflappable of African airlines, cancelling flights to Kinshasa for fear of being hijacked by hysterical soldiers, why was the president refusing to leave?

'Maybe he is hoping that if he waits long enough the rebel alliance will fall apart,' speculated a Zairean banker. 'Or maybe he's hoping Kabila will realise taking Kinshasa by force would be a very bloody affair. Or maybe he thinks he can simply buy Kabila off, as he has bought off so many others. One of these days he's going to wake up, call for one of his aides and discover he's the only mouvancier left in town.'

To think that Mobutu was hanging on to power for its own sake, to equate him with the generals clutching their posts for a few million dollars more, was to misunderstand what made him tick, insisted Nzanga. 'It wasn't a question of power. It was a question of the country's future. He was still trying to find a solution. Not to have bloodshed in Kinshasa, not to have someone like Kabila in control. Nobody wanted to understand the nature of his fight.'

It was at this point, one senses, that the lack of reliable intelligence, the decades of toadying by aides who told Mobutu only what he wanted to hear, finally took their toll. Long after the succession of army defeats made it clear that there was nothing to stop the AFDL, Mobutu still, to the bafflement of Western powers and his own population, seemed to believe he had cards to play. Nursing a fatal cancer, he knew his rule was coming to an end. But he was holding out for a deal which would allow him to exit gradually and with dignity, remaining perhaps as titular head of state while a transitional government took over administration and prepared the long-promised elections.

It was a scenario unacceptable to Kabila, who had said the only thing to discuss was Mobutu's departure, and one the president was in no position to enforce: but procrastination had served Mobutu well in the past. The longer he hung on, he calculated, the greater chance some former friend would come to the rescue. 'Emissaries were being sent out left, right and centre to try and get someone to do the fighting the Zaireans wouldn't do,' said former US ambassador Daniel Simpson. 'They were reaching out in every direction they could.'

Mobutu must have looked longingly across the river to Brazzaville, where a crack force of 2,000 Western troops had gathered. Ostensibly they were there to evacuate expatriates from Kinshasa if all hell broke loose, but he knew how such missions could end up rescuing a faltering regime. With this in mind, the French embassy in Kinshasa kept raising false alarms, pushing for intervention, only to be slapped down by Western chancelleries who realised the implications.

It was with the mission of shattering lingering presidential illusions that the Americans came to the villa on 29 April 1997 bent on engineering what Bill Richardson, the troubleshooter Bill Clinton entrusted with this delicate task, called a 'soft landing' for the rebels. For Mobutu's entourage, unwilling to accept the embarrassing truth about their army, US readiness to act as intermediaries proved what they had suspected all along. Washington, their former friend, was the secret weapon that explained the AFDL's extraordinary success.

The delegation entrusted with presenting what had been flagged as 'Mobutu's last chance', had been carefully picked to include representatives from the CIA, State Department and National Security Council. 'Mobutu had this trick of playing one side off against each other. We wanted to make it 100 per cent clear this was the US government position,' remembered Simpson, who did the translating. The man they met, seated on a throne, surrounded by his family and aides, was a shadow of his former self. The cancer had advanced and Mobutu, they saw, was now having difficulties walking, sitting and standing. He had become so cut off from events outside, the visitors found themselves in the bizarre position of giving him a military update, assuring him that Kenge, the last major town on the route into Kinshasa, had indeed fallen to the rebels, whatever his generals might be telling him about a DSP recapture.

But they did not allow his fragility to dilute the force with which they delivered their message. Arguing that the crisis had now reached an irreversible stage, they appealed to the president to step down 'with honour and dignity' while there was still time. Nzanga remembered that the team was sweating with nerves as they handed over a letter from Clinton in which the US leader urged Mobutu to meet Kabila and appoint a government team to negotiate a transfer of power. 'Richardson spelt it out in words of one syllable,' recalled Simpson. 'It was a very stark presentation. It was heavy-going, as you can imagine. This was a guy who had worked with the US since the 1950s and he was being told: "You'll be dragged through the streets. These things could happen to you and we are not going to stop them." '

The Americans then called a halt and the family went off into a huddle. Mobutu, Simpson remembered, was concerned that the rebel forces, also heading for Gbadolite, might desecrate his mother's tomb. The president made one brief, limp attempt to remind his guests of past loyalties, to stir old Cold War embers back into life. 'He said: "If you want to stop this, you can call in your troops." ' But history had moved on. 'We made it clear he wasn't going to get that.'

Exactly what deal Richardson's team offered in exchange for

Mobutu's voluntary departure remains a topic of dispute. In his memoirs Ngbanda, who was present at the meeting, claims that in return for Mobutu's withdrawal, the Americans said they would guarantee the safety of the president and his family, promised the MPR would be allowed to operate in the new political landscape—a way of saying his political legacy would not be completely obliterated—and added: 'We will ensure that your possessions, both inside and outside the country, go untouched.' The pitch, he maintains, was made orally, and certainly no trace of it appears in the letter from Clinton.

Even taking into account the understandable desire to win over their interlocutor at such a key moment by making promises that could later be quietly rescinded, this last offer, made to a man reputed to have salted away a fortune in stolen state assets, was one the US had no right to make and no legal jurisdiction to enforce. Ambassador Simpson simply denies it was ever voiced. 'We guaranteed Mobutu's personal safety, that's all. The rest comes from the Terminator.'

He is challenged by Nzanga, who was also present. While avoiding going into details, he confirms the account given by the Terminator, his arch enemy, as substantially correct. In assuming that an appeal to Mobutu's materialism would make a difference, he says, the Americans showed a crass misunderstanding of his father's thought processes. Having logged the decades of stolen profits, the riches of Gbadolite, the Americans must have thought they were touching the nub of the matter, raising an issue close to Mobutu's heart. But for the former army sergeant, wealth had always constituted a method, a tool for getting what he wanted, never an end in itself. 'That was never his concern. What bothered him was having to hand over power to a bandit, a man who had run smuggling rings and taken Westerners hostage. That was too much to swallow. The way they should have dealt with this problem was to talk to him as the father of the nation, and not deal with him as you would deal with a businessman, because that was an insult and he would never accept that, never.'

Due to fly on to meet Kabila the following day, Richardson

demanded a swift answer with what must have come across to the Zaireans as unconscionable American arrogance. Seething, but aware he was being presented with an ultimatum, not an exchange of views, Mobutu indicated that he would accept the deal, but needed time to put his agreement in words. Feverishly, Ngbanda and Vundwawe drew up the letter for Clinton that would end Mobutu's rule. But when the American team returned for their second meeting, the letter had vanished. Without notifying them, Mobutu had scrapped the resignation offer, agreeing only to a face-to-face meeting with Kabila.

For Ngbanda this was the final proof of the Mobutu family's interference in affairs it did not understand. However, another explanation is just as plausible. Mobutu was not used to being dictated to and the blunt language of the US team had stuck in his gullet. No decision at all seemed preferable to one that involved bowing to the nation he felt had masterminded his overthrow. 'A man like him could never have signed a surrender,' said Nzanga. 'He did not want to quit. He would rather have died.'

The meeting between Kabila and Mobutu took place on 4 May, after interminable wrangling over venue. With Mobutu refusing to fly to South Africa on the grounds of ill-health and Kabila ruling out either Gabon or Congo-Brazzaville for fear of a French-masterminded assassination attempt, South Africa came up with a compromise solution in the form of SAS *Outeniqua*, a navy vessel which was redirected to the port town of Pointe Noire.

The occasion had its share of black comedy, threatening to collapse entirely during the five hours officials spent debating how to get the president, too weak to climb the steep metal steps from quayside to deck, onto the vessel without making him look ridiculous. The president's doctor had warned against the effect of helicopter vibrations. Winching him aboard like a piece of cargo was deemed unacceptable. Finally, a makeshift ramp was built and Mobutu was driven aboard the *Outeniqua* in his bullet-proof limousine.

It was all to no avail. The two men's positions were too far apart for them to have anything to talk about. Photographs of the summit show a gaunt Mobutu and a chubby Kabila beaming for the cameras while still managing to look thoroughly ill at ease. The rebel chief is gazing at the ceiling, the sky, anywhere but into the eyes of his adversary, the only way, he had been told, to avert the spell Mobutu, that practitioner of black magic, would undoubtedly try to cast upon him.

With no deal reached, a grim scenario loomed—a showdown between the AFDL, UNITA and 10,000 elite Zairean troops inside Kinshasa that would probably trigger a complete breakdown in law and order and enormous loss of life. It was to avoid this outcome that Ambassador Simpson set about cultivating the generals, focusing in particular on General Donat Mahele, the unhappy head of Zaire's armed forces.

General Mahele was a career professional, a nationalist who won his spurs in the Shaba wars and had been appointed chief of staff in 1991. He had achieved huge popularity amongst ordinary Zaireans for ordering his troops to open fire on looting soldiers during the 'pillages' that devastated Kinshasa. But his forthright action made him no friends amongst the other generals. Worse, while he came from the Equateur region, he was not a Ngbandi. Sacked, he had gone into semi-retirement on his plantation until the 1996 crisis forced Mobutu to call upon his services again.

Mahele was too bright not to register swiftly that he could not win this war. Facing the prospect of humiliating defeat, nursing distant political ambitions of his own, he was receptive to any suggestions the Americans had to make. 'We began to talk about the shape of a soft landing, a situation in which Kabila's troops would come in not fighting and Mobutu's troops would maintain order without firing and hand over power,' said Simpson. 'Mahele said he could not stop the Alliance and that he was very interested.' The two men discussed the advantages of Mahele establishing direct contact with Kabila, an act that, in a nation at war, constituted high treason. 'Mahele said: "I have to think about it." I said: "You'd better, as you could wind up dead." '

Accompanying Richardson to Lubumbashi, the new AFDL base, Simpson made a note of Kabila's satellite telephone number and arranged for a call to be put through to his residence when he was back in Kinshasa. Mahele was invited around to receive it on 13 May. Aware that a botched 'soft landing', in which the rebels came into Kinshasa without shooting and the FAZ opened fire, would be far worse than none at all, Simpson asked Mahele: 'Do you really want to do this?' 'Yes,' replied the general. Knowing that Mobutu's intelligence services might be listening in, the ambassador was careful to avoid using any names once he picked up the receiver. 'I just said: "I'd like to introduce someone who wants to talk to you," handed Mahele the phone and walked out of the room. They spoke for about half an hour. There was a second call a couple of days later when they really nailed down procedures.'

But Kinshasa is not a city in which secrets are ever kept for long. General Mahele's visits to the ambassador's residence did not go unnoticed by his subordinates or his peers. With the US regarded as the steel in the AFDL glove, the inference was obvious: Mahele was in league with the rebels.

For the Inseparable Four, the revelation did not come as an enormous shock. The various generals—the recently appointed prime minister Likulia Bolongo in particular—had also been in talks with Kinshasa's foreign embassies, presenting themselves as Mobutu's natural successors, perfectly placed to negotiate with the rebels once the dinosaur was out of the picture. They did not appreciate being pipped to the post. So when, on 15 May, the generals asked for an urgent meeting with Mobutu, the Inseparable Four played Mahele a dirty trick, the last and most vicious in their long history of rivalry.

The day before had seen a diplomatic débâcle. With international encouragement, Mobutu had made a second, laborious trip to the SAS *Outeniqua*. What he hoped to achieve is not clear, but this time Kabila did not even bother to turn up, too confident now to mind that his no-show represented a slap in the face for both Mobutu and South African President Nelson Mandela, who had flown over to act as peace-maker.

In Mobutu's absence, the generals agreed amongst themselves the time for candour had arrived. A group would formally notify

Mobutu of what had been blindingly obvious to Kinshasa's residents for months: they could neither defend the city nor guarantee his safety. The message, it was assumed, would prompt Mobutu to announce his retirement. However, once before the president, the generals turned silent. They left it to Mahele to deliver the news that signalled all hope was at an end, then feigned surprise, casting the general in the role of turncoat. The explanation for the important television announcement that never was, it was the equivalent of handing Mahele the black spot.

Mobutu emerged from the meeting incandescent with rage, having finally, belatedly registered how truly exposed he was. Likulia lost no time in telling his subordinates he had replaced Mahele as head of the armed forces. If he had been on his own, Nzanga believes, Mobutu would have stayed in Kinshasa to accept his fate. But he did not want those who cared most for him to pay with their lives for their devotion. He allowed himself to be nagged and bullied by the family into a limousine and they set off on their last trip to Ndjili airport.

One story goes that as the presidential couple climbed the steps of the aircraft that would take them to Gbadolite, where the generals had assured them they would be safe, Bobi Ladawa turned and said to General Mahele: 'Donat, we know what you did. After all we have done for you, this is how you thank Papa.' Within twenty-four hours Mahele was dead, the victim of what most observers believe was an orchestrated plot. At some point in the afternoon of the following day, an army official rang Mahele to tell him the DSP was rioting in Camp Tsha Tshi. Abandoned by General Nzimbi, who had departed notwithstanding his morale-boosting speech to the troops, the DSP was going berserk.

It was a warning which would have prompted many a lesser man to head in the opposite direction. But Mahele had already drawn up a speech he planned to read over the radio the following morning, recommending a general surrender, and had arranged to fly to the Zambian capital Lusaka to formally recognise Kabila's authority. A DSP mutiny now would sabotage his plans. He drove immediately up the hill to the barracks, where he was confronted by a crowd of furious soldiers. As he tried to quieten a mob accusing him of selling out, someone walked up and shot him in the head, blowing out his brains.

The finger of blame has long been pointed at Kongulu Mobutu, the brutal DSP captain, who stayed behind in Kinshasa to hunt down those responsible for his father's overthrow and would certainly have been well aware of Mahele's treason. But Nzanga rejected the charge against his brother. 'Why Kongulu? Put yourself in the shoes of the Zairean soldiers. Nzimbi had gone, they had learned that somehow Mahele was connected to the rebellion. There were thousands of soldiers with thousands of reasons to hate Mahele.'

One of the two parties to the soft landing was dead. As the news slowly filtered out, journalists kept ringing Simpson for confirmation of the cataclysmic news. But aware that the rebels were now within hours of reaching Kinshasa, the ambassador trod water. 'I kept saying: "Have you seen the body?" ' Feeling more than a twinge of responsibility for Mahele's death, he was desperately hoping that events on the ground would acquire their own irreversible momentum before the news was confirmed.

Indeed, the rebels were marching in along the railway and, notwithstanding Mahele's disappearance from the scene, the FAZ was performing their traditional wardrobe change as the hardliners headed across the river. They were leaving so quickly, in fact, that at one stage the British and American embassies became concerned that a dangerous power vacuum was opening up and actually called Kabila to urge him to speed up the exhausted rebel force's advance. 'The troops were pouring in,' remembered Simpson. 'By the time Washington rang me to say: "Mahele is dead, does that mean the soft landing is off?" I could say: "No. It doesn't matter. It's done." '

As for the Mobutu family, they had come face to face with the extent to which the generals had betrayed them on leaving Kinshasa. Taking off for Gbadolite, a DSP colonel travelling with them told the pilot not to head out from Ndjili airport in the normal direction, heading east over the marshes. On his instructions, the plane headed west instead, then looped around in a wide circle. The reason for this manoeuvre? Before his departure, the colonel claimed, General Nzimbi had arranged for a jeep with a surface-to-air missile to be stationed under the route a Gbadolite-bound aircraft would normally take. In a rerun of the downing of the jet carrying Rwanda and Burundi's presidents, Mobutu's own cousin planned to wipe out the family in one clean sweep, then blame the crime conveniently on the AFDL. 'I'll never forgive Nzimbi. Never,' said Nzanga.

The treachery did not stop there. Arriving in Gbadolite, the family realised they were no safer there than in Kinshasa. The years of neglect, all those unpaid salaries of the local DSP, now came home to roost. Retreating before the rebels, the loyal elite was on the brink of turning against the Mobutu clan. It was time to go. But a practical problem presented itself. The family had dispatched their plane back to Kinshasa to pick up the remaining members of the entourage. And so the family that once hired Concordes without a second thought was reduced to borrowing a vast Russian cargo plane owned by UNITA head Jonas Savimbi. Braving what Nzanga remembers as 'a very, very hostile group of DSP' at the airport, the president, his family and a handful of loyal soldiers piled helter-skelter into the Ilyushin and took off.

Just in time. As the plane gathered height, the DSP men opened fire, their bullets ripping into the bodywork of the unpressurised aircraft, whose simple design probably saved the passengers' lives. 'God's hand was on us that day,' remembers Nzanga. 'It was lucky it was a Russian plane. If it had been a Boeing it would have exploded. From that moment on, my faith strengthened. I could never have seen my son again, my daughter would never have existed. For the first time in my life I stared death in the face.'

They landed on friendly soil in the West African state of Togo, to be met by an astonished Ngbanda and General Ilunga, who had gone to Lomé in a last-ditch attempt to try and activate Nigerian and Togolese promises of military help. Penetrating inside the interior of the bullet-pocked Ilyushin, Ngbanda helped his president extricate himself from the Mercedes that had been driven into the hold and scramble through a heap of suitcases and personal belongings. The first couple made their way shakily down the steps, to where an official motorcade had been hurriedly mustered by a surprised Togolese prime minister. Before allowing the car to drive off, Mobutu lowered the passenger window and addressed his security aide in a voice that was barely audible. 'Ngbanda, do you realise that even Nzimbi abandoned and betrayed me?' the president said in disbelief. Then he burst into tears.

In most diseases, psychological factors play a role in accelerating or delaying the illness's advance. Having lost his nation, with the fate of a people no longer resting on his shoulders and abandoned by those he trusted, Mobutu gave up his struggle against prostate cancer. Five days after the dramatic flight to Togo, the family began a new life as guests of King Hassan of Morocco. If the monarch remained a loyal friend to the end, Mobutu's exile was accompanied by a concerted washing of hands by former Western allies, timing to the second his transition from Helmsman to has-been. In his Rabat residence, Mobutu swiftly succumbed to bitterness and depression. 'It was very, very difficult. He began thinking about all the people he trusted who had abandoned him. And seeing the country he'd fought for all his life ending up in such a mess hurt him,' said Nzanga. 'At one point he talked a lot about the situation, then he stopped mentioning it and silence set in. He had begun to internalise it and his physical state deteriorated.'

In September 1997, less than four months after fleeing Kinshasa, Mobutu died. Far from his beloved forests and vast river, a sick leopard fading away in the arid dryness of Morocco, he had lived just long enough to see his achievements discredited, his reputation besmirched, his name vilified. There was a quiet funeral in Rabat's Christian cemetery. Ngbanda, who flew in for the event, was amid the group of former aides, personal doctors and bodyguards who stood at the grave after the family had withdrawn. Stricken by a sense of collective guilt, military and civilian alike sobbed aloud, begging their late master for forgiveness.

Nothing could have been more merciless than this interment in exile. In an African society only recently touched by urbanisation, where the spirits of the dead vie with the living for respect, burial outside the land of one's ancestors is worse than unnatural. For the man who had created the very nation of Zaire, with all its warts and blemishes, it could never constitute a laying to rest.

Crueller still was the final glimpse the Congolese public caught of the leader who so dominated their existence for thirty-two years. When the first AFDL rebels warily entered Camp Tsha Tshi, where DSP soldiers too old or low down in the hierarchy to bother fleeing were waiting to surrender, they found the looters had beaten them to it.

On the lawns and patio where a haggard president had allowed himself to be badgered by the press one last time, paperwork lay scattered. There were used cheque books for numbered presidential accounts in Belgium, scattered correspondence and, most poignantly, a letter sent to Zaire's consul in France by a Monaco-based intermediary purporting to have had contacts with the South African mercenary outfit Executive Outcomes. Dated 15 March, the day the city of Kisangani fell to the AFDL, the letter suggested an urgent meeting between Mobutu and Executive Outcomes bosses 'so that Executive Outcomes can do for Zaire what it has already done for Sierra Leone and Angola'. Did Executive Outcomes set too high a price for its intervention? Had the Zaireans dithered until it was too late? None of it mattered now.

In the presidential couple's villa, where Richardson and his team had delivered their ultimatum, the looters had been through all the rooms, opening every drawer, ripping apart every package, removing anything that could be easily carried. In the stainless steel kitchen, they had stolen the taps. From the headless pipes, water was pouring in a steady stream across the ground floor. On the spreading lake bobbed the contents of the boxes left behind in the family's rush, evidence of a dying man's physical collapse exposed for the world to see. Hundreds of adult incontinence nappies lay four or five layers thick, floating on the water.

If no one is a hero to his valet, every man is an object of ridicule to the burglar rifling through his bathroom. The image is hard to erase: a group of ragged looters, practitioners *par excellence* of Article 15, laughing and jeering as they throw the symbols of their toppled president's humiliation into the air.

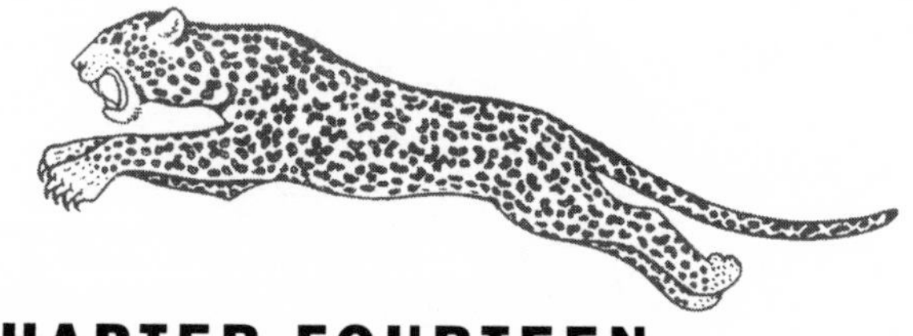

CHAPTER FOURTEEN

Ill-gotten gains

'The elephant is dead, but its tusks and hair remain.'

Bas-Congo proverb

For an African millionaire who liked to be near water, the northern banks of Lake Leman might have seemed a natural place to invest in real estate. From these vertiginous slopes, divided into a patchwork of vineyards, the view across the dark blue expanse to the brooding Alpine range separating Switzerland from France is Byronic in its drama.

'L'homme du fleuve', the man of the river, was not insensible to such charms. While Mobutu was being treated for the cancer that killed him, he caused a bit of a stir by booking two entire floors of Lausanne's appropriately named Beau Rivage Hotel. From here he could look directly out onto the water, across whose surface the currents trace silvery snail's trails.

But when it came to buying his own place, the president turned his back on Leman and headed higher up into the plateaux of the Canton de Vaud, where the gradient levels out, the road undulates gently through apple orchards and the locals enjoy a reputation for rustic ponderousness. He chose the village of Savigny, some 40 kilometres from Lausanne. Popular with celebrities because of a microclimate that remains dry and clear while a thick blanket of winter fog settles on the valleys, it was a convenient vantage point from which to monitor Yoshad, the 'import-export' business Kongulu Mobutu had opened in the town of Martigny, a perfect cover for transfers of money from Zaire to Europe. More important for his father, however, was the fact that Les Miguettes, the converted farmhouse he bought on Savigny's outskirts, was entirely hidden from public view by a thick screen of firs. The trees prevented any appreciation of the panorama, thereby removing, some might think, much of the point of living here. But personal security had become the priority in the latter part of Mobutu's life.

The five-hectare, thirty-room piece of real estate, valued, it is said, at $5 million, is easy to locate. On a country road where the Swiss flag flaps proudly over carefully clipped hedges, it is the only property that reeks of neglect. Beyond the wrought-iron railing running the length of the grounds, a neat lawn has grown wild. There are cobwebs on the large metal gates, left carelessly ajar despite an 'Entrée Interdite' sign. An intercom flickers into life just long enough for a suspicious Congolese voice—belonging to one of two homesick sentinels—to inform the curious this is strictly private property.

Nearly three years after Mobutu's flight into exile, Les Miguettes is stuck in limbo. Sequestered under a legal order that has been renewed, challenged by Mobutu's family, and renewed again, it can neither be used nor sold, belonging to neither the Mobutus, the new Kinshasa government nor the Swiss state. With its fate hanging in the balance, this prime piece of real estate looks doomed to slow disintegration. It is just one of the many properties in Mobutu's European real estate portfolio whose state of abandon attests to the abject failure of the institution established with much fanfare in Kinshasa in the months that followed Kabila's takeover: the quaintly baptised Office of Ill-Gotten Gains (OBMA), symbol of a new administrative drive to break with the past.

Nestling in the quiet lanes behind the Hotel Intercontinental, its windows opening onto nodding palm fronds and fragrant magnolia, the white-washed villa converted into OBMA's headquarters was a gift from Mobutu to the late Marie Antoinette. As such, it is almost certainly itself an ill-gotten gain. A fitting venue, then, for an organisation the AFDL promised would eliminate the tainted instincts the Congolese had inherited from the Leopard.

With a parking lot full of repossessed cars as an example of what it could do and its salons milling with chit-waving supplicants, OBMA set about a sweeping anticorruption campaign in mid-1997. 'We have a historical duty, to show generations to come just why we are underdeveloped,' explained Jean-Baptiste Mulemba, the nightclub owner turned guerrilla fighter, turned first OBMA director, a man whose commitment to the job in hand seemed to verge on the fanatical. A key aim was to secure the return of $14 billion that Mobutu had allegedly salted away in Swiss bank accounts, foreign corporations and luxury real estate.

European and American lawyers specialising in the stratagems Third World dictators used to conceal capital flight were lending their expertise, he claimed. He clearly envisioned a repeat of the legal campaign launched to reimburse the Philippines for Ferdinand Marcos's depredations. 'We'd love to recover all Mobutu's money. But recapturing even 60 per cent wouldn't be bad. If we can evaluate with our own methods the size of that fortune, I don't think any country will refuse to hand over the funds in the end.'

In the meantime, there was plenty of work OBMA could carry out unaided at home, tracking down government properties taken over by mouvanciers, ministry cars appropriated, bank loans never repaid. To squawks of outrage, OBMA's agents toured the villas of Binza and knocked on the doors of Kinshasa businesses. They challenged residents for proof of ownership, demanded explanations for the mysterious 'commissions' registered in company accounts and queried tax concessions granted in exchange for seats on the board. Flats, homes and shops were seized as OBMA officials struggled to prepare the paperwork required for hundreds of planned court cases. 'We will spend as long as it takes to expose what went on: ten years, twenty years, a century if necessary,' swore Mulemba. If OBMA aimed to become Congo's equivalent of the Truth Commission in South Africa—an institution spearheading the moral reawakening of a population—it would differ in one key respect: no amnesties were on offer. 'There will be no pardons,' he barked. 'If we forgive, those who come tomorrow will also steal. Once is enough. There won't be a second time in our country.'

Confronted with the AFDL's legal and moral crusade, the silence from France, Mobutu's most faithful Western friend, was deafening. But stricken, perhaps, with retrospective guilt, other countries signalled their readiness to co-operate. They blocked the assets of over eighty exiled mouvanciers and sequestered presidential properties. In Belgium, where many of the Big Vegetables held their accounts, enthusiasm was high enough for a pipe-smoking magistrate to fly to Kinshasa with his legal team in December 1997 to garner evidence of misappropriation.

No country was swifter in offering to help than Switzerland. Under media fire for their role as bankers to the world's undesirables, from the Nazis to Russia's mobsters, the Swiss had actually taken the risk of ordering Mobutu's assets to be frozen the day before the president was toppled. Determined to prove the efficacy of new legislation curbing the banking secrecy that drew a third of the world's offshore funds to Switzerland, they ordered the country's 406 banks to search for accounts in Mobutu's name. The outcome was anticlimactic. Instead of the $8 billion nest egg denounced by the AFDL's new justice minister, the Swiss banks located just six million Swiss francs ($4 million) belonging to Mobutu.

By late 1999, even that disappointing sum showed no signs of being repatriated to a needy Democratic Congo. Letters sent to Kinshasa by the Swiss police, keen to expedite the matter, received no response. 'We need more information and it has never come from Kinshasa,' explained Folco Galli, spokesman for the Federal Police Office. 'They must at least show us there is some link between these assets and supposed crimes, a suspicion, if not actual proof, for the dossier to go any further.'

It was the same story in Belgium. After a year spent waiting in vain for Kinshasa to prove the mouvanciers' frozen bank accounts contained stolen proceeds, the Belgian authorities reluctantly lifted the measure. 'Oh, I did have some problems at the start, but now everything has been sorted out,' was the refrain from the Congolese exiles who liked to take tea in Brussels' Conrad Hotel, a venue chosen, perhaps, for its passing resemblance to Kinshasa's Hotel Intercontinental. 'They couldn't prove anything.'

The sudden waning of government interest in the issue of Mobutu's stolen billions reflected, I discovered on returning to Kinshasa, dramatic changes back in Congo.

Fifteen months after the end of the AFDL campaign, a second war had broken out. Treading the same path as Mobutu, Kabila had made the mistake of underestimating how fiercely Rwanda and Uganda resented the presence in east Zaire of the Hutu interahamwe. In an extraordinary turnaround, he had welcomed Rwanda's genocidal killers—the very men his forces once tried to wipe out—into a new Congolese army, allies against the Tutsis he now suspected of plotting against him. Sure enough, Banyamulenge forces in the east mutinied, just as Rwandan units on loan from Kigali launched an abortive *coup* attempt in Kinshasa. The Tutsi ministers in Kabila's government fled to Goma, where they denounced the president as a tribalist and took nominal leadership of a new rebel movement. What the residents of east Zaire cynically referred to as 'our second so-called liberation' had begun, with exiled mouvanciers and generals providing funding and equipment.

Summoning Zimbabwe and Angola to his rescue, Kabila fractured the coalition of neighbouring powers that destroyed Mobutu. But the scenario dreaded by the Chester Crockers and Larry Devlins of this world had finally come about: Congo was effectively divided in two as greedy neighbours plundered its mineral spoils. Unable to rely on the useless Congolese army, Kabila exchanged Zimbabwean protection for a majority share in Gécamines, while squabbling rebel factions in the east traded gold and diamonds for Ugandan and Rwandan military support. Congo had become a shifting, unsteady core radiating instability across the continent.

So much had changed, yet so much seemed curiously the same.

With the new war costing it dear and Western donors cold-shouldering Kabila for his human rights record, plans for national reconstruction had been scrapped. The salaries of the civil service and army were arriving late or not at all. Fuel was being rationed and queues of cars radiated for kilometres from every pumping station. The administration was so desperate for cash, even practitioners of Article 15, players in the plucky informal sector that had succeeded in escaping Mobutu's scrutiny, were now being taxed.

Afraid of assassination, the president rarely appeared in public and had entrusted his safety to a military elite from his home province. Only this time the special units who picked up government officials and former politicians too free with their opinions—Cleophas Kamitatu was one of many such—came from Katanga, not Equateur. Once hailed as belonging to a 'new breed' of reform-minded African leaders, Kabila, it was becoming clear, did not really believe in consensus rule. He rarely held cabinet meetings, reshuffled posts repeatedly and entrusted important decisions to relatives appointed to key positions.

The independence of the central bank had been quietly rescinded, with real power resting with a committee stuffed with sympathetic ministers. Citing the war effort as justification, another president unable to differentiate between public and private purse dipped at will into the coffers of the central bank, which printed money frenetically to cover the gap. 'There is absolutely no difference between the management of the central bank under Mobutu and under Kabila,' confessed a depressed government economist.

In another move redolent with *déjà vu*, Kabila had outlawed political parties and launched Popular People's Committees (CPP), building blocs of a grassroots movement intended to embrace every citizen. The CPP's initials, joked Kinshasa's residents, who had seen it all before with the MPR, really stood for 'C'est Pas Possible'—'It's Not Possible'.

Most hilarious of all, in their eyes, was the reappointment of Sakombi Inongo, the man who had built Mobutu's personality cult.

Originator of the television broadcast showing 'Papa' emerging from the heavens, Sakombi was now marketing 'Mzee'—the respectful Swahili word for 'elder'. His handiwork was plastered around Kinshasa: huge posters of the overweight Kabila, under the headline 'Here is the man we needed'. The Intercontinental's executives abroad clearly did not share this view. The chain had finally grown tired of the challenge posed by doing business with a debt-ridden Kinshasa government and had formally severed their links with the hotel once so beloved of the Big Vegetables.

While doyens of the old system were returning, those who had proclaimed its overthrow were departing. Superficially, OBMA seemed intact. The magnolia still blossomed in the courtyard, the waiting rooms still buzzed with supplicants. But Mulemba the zealot was long gone. He had been thrown into jail, then released, after an audit suggested those investigating the alleged embezzlers had been helping themselves in the process. The second director, a Banyamulenge, had held the job only a couple of months when the Tutsi revolt broke out and he fled to join the rebels. I spent weeks courting the third. On the day we were due to meet he called me on my Telecel to cancel. 'Haven't you heard the news? It seems I've just been suspended.' Yet another audit showing missing funds, yet another scandal.

The rapid turnover in directors reflected the extent to which a new elite lusted after what was viewed as a ripe source of freebies. For far from lancing the Congolese boil of corruption and returning stolen assets to the people, OBMA had become a perfect cover for the seizure of jeeps, Mercedes, flats, villas and businesses coveted by a fresh array of Big Vegetables, whose very newness left them with more ground to cover in the race to self-enrichment.

'The problem begins at ministerial level,' said one OBMA official, blessed with the limpid vision possessed by so many Congolese, the ability to offer a merciless assessment of his society's failings without coming up with any idea of how to correct them. 'The nicest villas in Binza are always the ones to go.' Disputes were endless, he said, and almost always involved lodgings. One army commander, living in the villa once owned by one of Mobutu's former lieutenants, would envy a rival commander, who had moved into a larger home confiscated from a general. 'We're one big house-letting agency,' he sighed. So heated had the scramble for goodies between various factions become, OBMA agents had been arrested in Kinshasa and thrown into jail in Katanga and Kasai, where the local authorities wanted to be free to run their own confiscation scam. So much for flushing graft from the national psyche. 'You know, in the fight against Mobutu, not everyone shared the same objectives,' he mused. 'Some people wanted to change society. Some just wanted to replace him. It's the principle of "Ôte-toi de là, que je m'y mette" (Get out of the way, so I can take your place).'

What had become of the campaign to repatriate the exiles' stolen money, in particular the funds belonging to the most high-profile exile of them all, Mobutu? I asked him. The Congolese lawyer who initially dealt with the issue had never been paid, 'in order to discourage him'. The dossier itself had been taken out of OBMA's hands by the Justice Ministry, which was suspected of planning to close OBMA down altogether. 'There's a clear intention not to touch the assets of the mouvanciers,' he confessed.

Why such carelessness on the part of the institution that once vowed to clean out the Augean stables? Why was no one answering the letters sent by the Swiss and Belgian authorities? I found the answer in the padded office of Congo's friendly public prosecutor, reading what the latest incumbent assured me was the most recent paperwork drawn up in the Mobutu case. As the prosecutor fielded calls, I leafed through the four-page fax his successor had sent abroad in July 1997, the request for an asset freeze that at the time seemed to signal the start of something important, but now appeared to mark the high-point of an exercise in futility.

Mobutu, the head of state who had bled his country dry for thirty-two years, merited just five lines. His former aide Seti Yale, regarded by Congolese as one of the richest men in Africa, stood accused of misappropriating twenty-four state vehicles. Ngbanda Nzambo Ko Atumba, also known as the Terminator, came next, accused of stealing eighteen cars. Here too was Kamitatu, charged with appropriating two embassy buildings in Tokyo. Curiously, former Prime Minister Kengo Wa Dondo, reckoned to rival only Seti in his wealth, had not been judged worthy of a separate entry. He featured only on the list of eighty-three mouvanciers whose accounts should be blocked.

Given this half-hearted effort, indeed, it seemed remarkable that foreign judicial authorities had taken any action at all. Was this, er, it? I inquired politely. 'Well, we're leaving it to our friends abroad to fill out the details.' But, I said, I understood those officials were themselves waiting for Kinshasa to provide proof of criminal activity. A vicious circle seemed to be at work, with each side expecting the other to deliver. The prosecutor looked thoughtful. To be honest, he acknowledged, investigations had been disrupted from the start, 'disrupted almost voluntarily'. Too much time had been allowed to elapse. Telexes had been delayed, files sent to the Justice Ministry never returned. He was still awaiting official approval for a trip he hoped to undertake to Switzerland and Belgium one day to pursue the affair. 'There's a lack of enthusiasm on the part of the judicial system. Even the presidency doesn't seem to be interested.' He paused. 'At the moment, we are following another path,' he said, and it was clear the 'we' referred to Congo. 'Given the desire for everyone to get on well together, this has become a secondary issue.'

Comprehension dawned. I had heard reports of an official softening in stance towards former Mobutu cadres. The planned trials of over forty mouvanciers in Makala prison had been quietly dropped. Instead, each had been allowed to buy his way out of jail—in some cases with a $1 million contribution—in exchange for letters informing them the state now considered their cases closed. Feelers were being extended to the exiles in Brussels and Paris to try and persuade them to return home. Names familiar from the Mobutu era were circulating again. Jonas Mukamba, the former MIBA head, had moved back into his villa in Binza. Bemba Saolona, the magnate who had flourished under Mobutu, would shortly be made Economy Minister, part of an attempt to woo his rebel leader son.

With rebels holding half the country and a hostile West turning its back on Democratic Congo, an isolated government could not afford to be fussy about its friends. The mouvanciers the AFDL had sworn to bring to justice were now welcome if they brought cash contributions, the support of their ethnic groups and a modicum of experience back with them. So who wanted a legal investigation that would expose how soiled the partners in this new marriage of convenience really were? Principle had been sacrificed on the altar of expediency. Or, in the more tactful phrase of the public prosecutor: 'I think this is a dossier people want to forget in the name of national reconciliation.'

With investigations stalled by an embattled government making the cynical accommodations it needed to stay in power, the task of establishing the most basic premise of all—the actual size of Mobutu's stolen fortune—remained unfinished.

I have never been able to track down the fabled teams of European and American lawyers Mulemba told me had been hired as detectives by the Congolese state. I have spoken to US Treasury officials, Swiss bankers and policemen, legal experts from the International Monetary Fund and British law firms who specialise in this kind of work and badgered the Congolese authorities themselves—but all to no avail. The apparent absence of legal expertise has not stopped Congolese opposition parties, human rights activists and debt campaigners from claiming a vast fortune lies hidden abroad, still tapped by the president's extended family. Mobutu's name features near the top of a list which embraces Marcos and Bhutto, Noriega and Suharto: Third World leaders whose swollen assets serve as shameful indictments of a bankrupt Western policy based on indulgence and appeasement. For the campaigners, the paltry $4 million unearthed by the Swiss banks is a sign of either institutional hypocrisy or wilful naivety. An operator as wily and as well connected as Mobutu, they argue, would never make the basic mistake of depositing the bulk of his takings in his own name. 'They launch an electronic search for "Mobutu" or for "Bobi Ladawa". But you'd have to be crazy to keep accounts in those names!' scoffs Jean Ziegler, a writer and socialist Swiss MP who has dedicated his career to exposing the moral duplicity of the Swiss banking system. 'If you set up an off-shore society, which sets up a trust fund, which opens an account in a fictional name, then that's not going to show up on the computers.'

And Mobutu certainly avoided using his own name when it came to sending money out of the country, often opening accounts and establishing shareholdings in the names of trusted relatives, employees and friends. The practice, with all its inherent risks, was exposed after the death of Litho Maboti, Mobutu's uncle and one of his financial frontmen. Clearly not sharing his parent's sense of loyalty, the son asked a US lawyer to help him seize funds held in his father's name.

Yet the more you scratch at the huge numbers cited by Mobutu's critics, the less appears to lie behind them. The $14 billion estimate given by the incoming Justice Minister Celestin Lwangi in 1997, for example, is immediately suspicious because of the way it exactly mirrors the amount Zaire today owes its foreign creditors. This financial parallel can be traced all the way back to 1982, when German banker Erwin Blumenthal asked Nguz Karl i Bond, Zaire's disaffected prime minister, how much he thought Mobutu had in his foreign bank accounts. Nguz ventured an estimate of $4 billion, on a par with the Bank of Zaire's estimate of the country's foreign debt at the time.

In a system of personalised presidential control, the premier's post did not bring much power with it. It certainly did not entail knowledge of Mobutu's bank accounts, as Nguz himself admitted in the report. But being able to source such a figure back to a prime minister—a supposed figure of authority—was a godsend for those trying to draw international attention to Mobutu's iniquities. The president could cancel an entire country's foreign debt with one personal cheque! The parallel was a seductively simple way of conveying a point about African corruption and Western complicity. As interest payments went unpaid and Zaire's debt rose, the estimates of one man's personal hoard were similarly, automatically, upgraded: $4 billion, $6 billion, $8 billion, $14 billion. The authors of books on Zaire quoted the journalists, the journalists quoted the authors, the left-wing politicians quoted both, until eventually the sums became true by dint of sheer repetition.

The financial tandem is too neat to be convincing and it is noticeable that those who repeat the billion-dollar mantra rarely offer even the sketchiest of breakdowns. The most credible attempt has been made by Steve Askins and Carole Collins. Using leaked World Bank and IMF documents, they have logged some of the misplaced export earnings, the inflated presidential allowances, the nebulous spending on 'other goods and services' that gobbled up hundreds of millions of dollars each year.

But their reports provide only partial, tantalising glimpses. Wary of those who might one day come snooping, a man by nature impatient with financial detail, Mobutu left the most insubstantial of paper trails. Commands were issued verbally, withdrawals from the central bank or state enterprises nominally authorised by subordinates. 'He made sure his signature was never on a withdrawal slip,' remembers a political contemporary. 'Nothing could ever be traced back to him.' Deprived of access to bank statements and shareholding certificates, the two US researchers wisely skirt shy of offering any independent estimate for the size of the remaining presidential treasure trove.

But can Mobutu's aversion to paperwork alone explain this uniform failure to quantify or locate a missing stash? My own belief is that the crudeness of the methods Mobutu used to divert sums through the years, the vulgar ostentation of Gbadolite, the cellars of Laurent Perrier stacked in the tropical African heat, are all in danger of blinding us, just as they did the Richardson delegation, to the central truth about his style of rule.

Mobutu was certainly no ascetic, but money was always a method, an instrument, the most effective of the techniques available when it came to maintaining, extending and preserving his power.

'He never had the soul of a racketeer,' says son Nzanga. 'For him money was simply a means to an end, a way of getting what he wanted.' For once son and aide are in agreement. 'No one will ever lay their hands on a fortune, for the simple reason that it doesn't exist,' insists the Terminator. 'Yes, Mobutu liked to do things in style. He felt he was the head of a great country and could live the good life. A lot of money went through Mobutu's hands. But it went through his hands and didn't stay.'

It was a clientilist system that gobbled cash. There were bribes to be paid to Western businessmen, politicians and journalists, wages for the DSP, donations to foreign guerrilla groups, gifts to generals, governors and opposition politicians. Even fairly trivial sweeteners, such as the Mercedes, Peugeots and jeeps provided to new ministers and their deputies, mounted up when repeated ad nauseam. A World Bank economist once calculated that if each vehicle cost $40,000, a reasonable guess given Zaireans' love of flashy cars, every new government team spent on average $4.8 million on vehicles. Multiply that by the fifty-one government cabinets Mobutu appointed and at a rough guess, $250 million was spent on Zaire's state cars alone between 1965 and 1990.

Mobutu's theft, concurs Kim Jaycox, the World Bank's former Africa supremo, was a measure not of greed but of political weakness: he needed the money to remain head of one of Africa's largest, most fractious states. 'The country was kept together by the loyalty of the regional governors, who were essentially warlords. Mobutu was milking whatever cows he had and sending the money to these guys. It was a very, very expensive business. Which is why when people understood what was going on, they closed their eyes.' Seen in this light, the thieving becomes a crude form of pork-barrel politics as practised by a leader who, like many an African contemporary, never grasped the concept of state as distributor of national revenue. Those visits to the provinces, in which Mobutu promised a hospital here, a school there, and handed out brown envelope after brown envelope, ensured the loyalty of regional leaders did not wander and the unitary state survived.

The demands on his purse were huge. The US Treasury launched preliminary investigations into the matter when, in the early 1990s, Washington, Paris and Brussels briefly played with the idea of forcing political reform on Mobutu by freezing his foreign assets. The Treasury's assessment was that, in contrast with public perceptions, Mobutu in his last years had been outstripped by his generals—active in the diamond and oil trade—when it came to revenue-raising. 'When we tried to get a hold on what he had, we found to our surprise that Mobutu was having serious cash flow problems,' said one official. 'He was having problems paying his bills, maintaining his French properties and keeping his entourage happy. It suggested that his ability to plunder various state mechanisms had shrunk enormously as Gécamines and Miba had decayed. He had squandered huge amounts and not squirrelled it away as was supposed.'

The US Treasury came up with a figure of $40–45 million, an estimate that only involved real estate, the most visible and therefore quantifiable aspect of Mobutu's wealth. Interestingly, this is close to the sum Mobutu himself revealed to an interviewer when in 1988 he was asked the size of his fortune. Estimating it at under $50 million, he added: 'What is that after twenty-two years as head of state of such a big country?' By 1997, there were even signs Mobutu was running out of spending money. Unable to maintain even his personal jet, he was reported to have pillaged funds set aside for the elections to pay for the Kisangani mercenary force. With his own survival in the balance, it seems unlikely he would have stinted on this had the cash been easily to hand.

When the subject of Mobutu's fortune arises, Nzanga rolls his eyes to heaven. Since going into exile he has received a letter from the AFDL asking for the return of $8 billion ('Eight billion!' he snorts incredulously) and has fended off numerous overtures from enterprising Congolese claiming that as Mobutu's illegitimate children they deserve a share in his inheritance. He knows, better than anyone, the spell cast by the fabled billions. 'There's so much hearsay. But it's all supposition which began with the Blumenthal report. I'm no beggar, but I drive a ten-year-old car and I can't afford to go out and buy myself a plane. All we have is some real estate bought by Mobutu, which doesn't amount to a billion French francs. Yes, there are large sums in the accounts of important men. But it's not with us. Millionaires tend not to be very generous men, and my father was very generous.'

The legends of Mobutu's hoard will endure, fuelled by reports of roomfuls of gold ingots in Gambia, front companies in the Canary Isles, bank statements left lying in the debris of looted Gbadolite. One suspects that Nzanga and his siblings will never be forced to apply for social security, but I am willing to predict that there will be no miraculous discoveries of billion-dollar Mobutu stashes, Africa's equivalent of the Loch Ness monster. Congo's stolen fortune is not hidden, but on open display. It has been invested in the luxurious mansions of Uccle and Rhode St Genèse in Brussels, where exiled prime ministers rub shoulders with former heads of industry. It is to be found in the South African compounds where chauffeurs polish gleaming ranks of Mercedes and on the Rolexed wrists of mouvancier wives flourishing credit cards in Parisian designer shops. This is redistribution of wealth at its most inegalitarian, a trickle-down effect that moistened no more than the uppermost layer of a social hierarchy.

Most Congolese long ago accepted that the elite which rode on Mobutu's coat-tails will never be brought to justice. And now that their initial enthusiasm for a financial spring-cleaning has subsided, Western governments and banks are secretly relieved by Kinshasa's indifference to the issue. Not only because the proceeds of the country's diamonds and copper pad out their balance sheets, or because embarrassing light would be shed on the role they played in condoning massive capital flight. Given the growing reports of Kabila's own financial mismanagement, the increasing evidence he has learned much from Mobutu, who, after all, could hand over the money with a sense of a wrong righted, a job well done?

Does the task OBMA was meant to perform, but will never now

complete, still matter? There are many, like Daniel Simpson, the former US ambassador to Kinshasa, who regard it as a dangerous irrelevance for a country with far more pressing problems to tackle. 'It's a street without joy, the moral equivalent of digging for pirate treasure,' he says. Yet to track down the missing money, to decide how much was taken and where it went, who turned a blind eye and who took a cut, who deserves to be compensated and who should be punished, would be to start the process of breaking with a hopeless past and building a state forty-five million Congolese might be proud to inhabit.

Despite the Leopard's departure, there has been no renewal, no change in mentalities. Mobutu ruled thanks to the support of a mono-ethnic security force. So does Kabila. Mobutu plundered the central bank. So does Kabila. Mobutu destroyed the formal economy. Kabila has gone even further, choking off the informal economy. 'Kabila,' as one European politician astutely remarked, 'has simply replaced Mobutu with Mobutuism.'

The depressing accuracy of his observation came home to me one winter's evening in Brussels, when I watched Kabila being interviewed during a rare visit to Belgium. Aimed at improving tense relations between the new regime and the West, the trip had not proved a success. It had been nearly cancelled altogether when the arrest of Chilean dictator Augusto Pinochet in London suddenly raised the possibility of Kabila being served with an injunction for war crimes for his role in the massacre of Hutu refugees. Unsure of how to treat him, the Belgians had failed to provide military honours at the airport and King Albert had made a point of not shaking his hand. Pinned under the blazing television studio lights, Kabila radiated a kind of suppressed fury. Drops of sweat stood out on the huge bald head that seemed to blend seamlessly with his torso. But he nonetheless refused to remove his dark cashmere coat as he fielded questions with a broad smile of fake bonhomie. He had none of Mobutu's mastery of the situation, none of the you-know-and-I-know-but-watch-me-as-I-tell-you-the-most-massive-whopper humour. Instead, he sounded like what he was: a thug being forced for one brief moment to answer his critics. Asked about a minister's arrest, he said he was 'unaware'. Questioned about the lynching of Tutsis, he chuckled. It was not a nice performance.

Looking at the screen I realised with a slight shock that in his dark coat, buttoned up to the chin, with his neck swathed in a woollen scarf, Congo's new president looked exactly as though he was wearing an abacost, that symbol of Mobutu's rule. I was reminded of the moment in George Orwell's *Animal Farm* when, staring through the kitchen window, the animals watch their self-important new masters, the pigs, fraternising with the farmers, oppressors of old. 'The creatures outside looked from pig to man and from man to pig, and from pig to man again: but already it was impossible to say which was which.'

EPILOGUE

'The conquest of the earth, which mostly means the taking it away from those who have a different complexion or slightly flatter noses than ourselves, is not a pretty thing when you look into it too much.'

Heart of Darkness
—Joseph Conrad

Every week after Mass in Rabat's cathedral, the members of a Congolese family, dressed in their Sunday best, are in the habit of making their way to the city's Catholic cemetery. The cemetery, which dates back to the era of French rule in Morocco, lies in the popular quarter of Akkari and looks out over the Atlantic Ocean. It is already choked with the graves of thousands of French colonials who died in foreign service and Christian families seeking to bury their loved ones are obliged these days to look elsewhere for a resting place.

Yet room has nonetheless been found within the grounds for a modern tomb, built from Italian marble and holding space for six coffins. The care alloted this particular plot by the cemetery guardians, paid a small allowance to freshen up the flowers placed there, is another particularity suggesting its father-and-son occupants are a little out of the ordinary. Over photographs of the older man two inscriptions have been carved. 'Here lies President Joseph Désiré Mobutu, born 14 October 1930 in Lisala, died 7 September 1997, in Rabat' reads one. The other gives the identity the inventor of 'authenticity' would no doubt prefer to be remembered by: 'President Mobutu Sese Seko Kuku Ngbendu Wa Za Banga, marshal.'

It is here that Bobi Ladawa and her sister Kossia come with their children and grandchildren to pay tribute to their dead patriarch, Zaire's Leopard, and frightening son Kongulu, who shared his father's fate of dying in exile. On the surface at least, the two women appear to have adapted smoothly to life in Morocco. The vast retinue of security guards and doctors, cousins and in-laws, hairdressers and maids that originally followed Mobutu to Rabat has dispersed, to the quiet relief of their Moroccan hosts, and members of the immediate family have moved into flats and villas in the capital's most exclusive districts.

In keeping with her former role as 'mother of the nation', the former first lady keeps a benevolent eye on the sizeable Congolese community in Rabat, a favourite jumping off point for youngsters bent on building new lives in Europe. She shares with her late husband a sense of the duties incumbent on the tribal chief, and is respected for her many acts of charity towards the less fortunate. She has lost a great deal of the weight put on during the years of plenty and is careful to keep a low public profile, perhaps aware that with the death of King Hassan, her late husband's last African friend has gone, weakening the position of a family reliant on the Moroccan regime's continuing hospitality.

But in truth, the Mobutu children have found their surname less of a hindrance than they might have expected, now that the Leopard's death has removed the main source of potential political embarrassment. They travel freely between Morocco, Europe and the United States and find, in common with other Big Vegetables who fled abroad, that crossing borders has become ever easier as the world's amnesia towards Zaire—a country that no longer exists—grows.

Like all exiles, they try to keep busy. But essentially they are waiting: for the day they can return to Kinshasa without fear of being jailed, for the day they can accompany Mobutu's body on its last flight, back to his ancestral home. Even the president's fiercest Congolese critics now feel repatriation is in order. But while longing for just such an outcome, the family hesitates to trust assurances proffered by a new administration. 'Kabila can talk in the air as much as he likes, but it won't be under him that we repatriate my father's body,' says Nzanga. 'Certain preconditions have to be met. He was, after all, head of state for thirty years.'

Maybe only a state funeral for the man they called 'Papa', a ceremonial day of reckoning, could put an end to the condition of arrested development in which an orphaned Congo seems stuck.

Having ducked a confrontation with its past, the country cannot forge an alternative future. If the end of the Cold War has left Congo master of its fate, the country barely appears to have registered the change. Locked into the habits of paternalism, with its political landscape still dominated by personalities that first hit the headlines in the post-independence years, it is running on rails laid by a dead man, bereft of fresh ideas.

Those who can, get out. I have sat on flights next to some who failed to escape: handcuffed Congolese immigrants, flanked by French gendarmes, who wailed as they returned to the country of their birth. These were not political activists expecting to be tortured on arrival. They were economic migrants frustrated in that most basic of urges: to be able to aspire to something more than mere survival. Whatever bloody deeds were carried out on his orders, this will always constitute Mobutu's worst human rights violation: the destruction of an economy that quashed a generation's aspirations.

But most Congolese will be forced to stay and ride out events, as the country serves as stage for its neighbours' battles and the number of rebel groups without a cause climbs exponentially. 'Maybe things have to get worse in Democratic Congo before they get better,' speculates an ambassador. 'There are now so many young men with guns out there, I can't see things improving for ten, twenty, maybe thirty years.'

It is worryingly easy to imagine that, one day, Congo's predicament may become so bleak its citizens will actually wax nostalgic for Mobutu, just as under Mobutu they talked with fondness and selective amnesia of the ghastly colonial years. The prospect of a political party, coalescing around one of the surviving Mobutu sons maybe, which would revive the concept of authenticity while glossing over subsequent absurdities, makes one shudder. But similar about-turns are happening across Africa, and for identical reasons. Failing to understand the reasons behind a country's ruin makes repetition all too easy.

This is not to say that the post-Mobutu years have been short of memoirs, penned by the key players in the story. But the publications all share one basic characteristic. In each case, responsibility for the disaster that was the Congo has been smoothly shrugged off, to be shifted wholesale onto the dead man's shoulders.

Mobutu's very charisma, the larger-than-life personality that so overwhelmed all those who met him, encourages such scapegoating. The leopardskin cap, the magic cane, the growling voice all lent themselves to caricature, the image of a monster whose sins dwarfed those of lesser men. For mouvanciers and opposition politicians, journalists and diplomats, half in love with the Mobutu myth, it was always easier to rail against Big Man rule than confront a more unsettling reality. Accept that no one figure stood at the nation's controls in the last years of Zaire's existence and each aide, general, foreign minister and financier would have had to acknowledge their own contribution to the system.

Yes, Mobutu was brutal, ruthless and greedy. Possessed of the instincts of the neighbourhood thug, he knew only how to draw out the worst in those around him. Most disastrous was the fact that he lacked the imagination, the sustained vision required to build a coherent state from Belgium's uncertain inheritance. But if Mobutu traced a Kurtz-like trajectory from high ideals to febrile corruption, he did not pursue that itinerary alone, or unaided.

The phrase that lingers in my mind was voiced by Nzanga Mobutu. 'When history judges my father, it will judge in detail,' he said, meaning blame would eventually be apportioned precisely where it was due. Citing members of a presidential entourage who criticised in exile the former patron they once indulged, he added: 'I think a bit of *mea culpa* would be in order.' *Mea culpa*. Throughout my interviews, I had kept expecting to find signs of it, only to be constantly surprised by its failure to make an appearance. There was precious little from the Washington financiers who granted billions to a known thief, whose institutions will one day have to explain why the Congolese should be held accountable for loans made in bad faith. Even less from the US and French officials who, motivated by strategic reasons, decided with cool cynicism what was best for this most fragile of post-independence states.

There was none at all from the Congolese aides, ministers and generals who helped mould the dinosaur's policies, still adopting the 'I was only following orders' excuse judged insufficient at Nuremberg. And with the remarkable exception of Jules Marchal, the guilt-ridden retired Belgian diplomat, the colonial power that first sent Congo on its wayward course had nimbly succeeded in dismissing the very notion of blame.

To explore the roles they played—from the raids of the slave traders to the amputations carried out by the Force Publique and the wishful thinking of the World Bank—is to move from exasperation at a nation's fecklessness to wonderment that a population has come through it all with a sense of humour. It seems surprising not that Congo is as shambolic as it is today, but that its condition is not far worse.

The quality of negative excellence that so unnerves visitors to the Congo, the country's capacity to take the faults of any normal African state and pitch them one frequency higher, were nurtured by a brutal colonial past, followed by a unique level of meddling by the Western powers.

Now that the US, France and Belgium have distanced themselves, now that Mobutu is dead, the country has lost the last excuse for its predicament. A population that has set its sights little higher than survival has to learn to take responsibility for its own destiny. 'What do the French want of Congo?' François my driver would ask me when discussing his country's future, a question that alternated with the equally infuriating, if equally understandable 'What do the Americans want?' The question must now become 'What do the Congolese want?'

In London one grey day, when news agencies were reporting on yet another breakdown of a supposed ceasefire in the Congo—another chance for some *Heart of Darkness* similes—I stumbled across a 1956 collection of black-and-white photographs of Leopoldville on the shelves of my London library. Bound in light blue, with the title picked out in gold—the colours of both Kabila and the Belgian Congo—it was published to mark the seventy-fifth anniversary of the city's founding.

Flicking through the pages was like travelling to another world. Blessed with the glorious unselfconsciousness of a time when colonial shame seemed inconceivable, the author proudly paraded a port full of cranes, factories turning out cloth, a modern railway network and 'one of the only two gyrobus transport systems in the world'. Kinshasa, it was clear, had once been a veritable Milton Keynes. If it was only four years before independence, there was little sense of a nation being prepared to take its destiny into its own hands. Instead there were photos of Congolese obediently bowed over lathes, typewriters and microscopes as white tutors gave instructions. On one page a Belgian housewife taught local women how to run a kitchen, on another, black chefs in white aprons demonstrated their skill at producing Belgian patisserie.

Before and after . . . the photographs showed jungle bulldozed to form a city street, oxen making way for cars of the 1950s, a model Congolese family relaxing in a spotless lounge, sipping tea as they listened to the radio. But a vital chapter was missing. Now. That would reveal the wheel turning full circle: the jungle growing back through the potholed tarmac, running water tainted with sewage, neighbourhoods without electricity, walking replacing the car.

I knew these streets, these roundabouts, these buildings. But I had never seen them so tidy. Here was the high-rise building now converted into the Memling Hotel. But where were the streetsellers who usually gathered outside it, with their selections of cigarettes, boiled eggs and cola nuts? Where were the house-high piles of rubbish, the polio victims in their tricycles, the begging albinos blistering in the sun? Could this really be the same city?

Feeling lost in this unfamiliar world of order, symmetry and seemingly unquenchable hope, I pored over each photograph, looking for some hint of the chaos to come. And then, halfway through my perusal, I was pulled up short. There, on page 144, was a photograph of a policeman directing traffic on one of the boulevards. His uniform looked neat, his gauntlets were a spotless white. But looking closely at his face, I could swear he was wearing gold-rimmed, slanting sunglasses—pimp's sunglasses, sinister trademark of the secret policeman and presidential guard, the torturer possessed of arbitrary, undefined powers. Now there, in that tiny, telling detail, was the country I had come to know and love.

Glossary

The nation carved out of central Africa by Belgium's King Leopold was originally known as the Congo Free State. When Belgium took over the administration it was dubbed Belgian Congo—to distinguish it from French Congo across the river—and was known as Congo after independence. In 1971 the country, its river and its currency were all rebaptised Zaire by President Mobutu. When Laurent Kabila took over in 1997 he reverted to the names of the previous era. The rechristening has led to some confusion, with Congo and the Congolese often mistaken for their neighbours across the water. Congo-Brazzaville is another country entirely, and not the subject of this book.

Names under Belgians	*Names under Mobutu*	*Names under Kabila*
Congo	Zaire	Congo
Leopoldville	Kinshasa	Kinshasa
Stanleyville	Kisangani	Kisangani
Elizabethville	Lubumbashi	Lubumbashi
Bakwanga	Mbuji Mayi	Mbuji Mayi
Katanga	Shaba	Katanga

Names under Belgians	*Names under Mobutu*	*Names under Kabila*
Coquilhatville	Mbandaka	Mbandaka
Stanley Pool	Pool Malebo	Pool Malebo

President Kabila's Congo is a place where being a little too free with one's opinions can cause problems with the authorities. In the very few cases where individuals living in the country have voiced views that could conceivably trigger repercussions, I have changed their names.

AFDL Alliance of Democratic Forces for the Liberation of Congo-Zaire. The coalition of four rebel movements set up in east Zaire in 1996, which swore to bring down Mobutu. Laurent Kabila, originally the movement's spokesman, became its leader.

CNS Sovereign National Conference. First convened in August 1991, this was a vast talking shop embracing political parties and representatives of Zairean civil society with a mandate to pave the way from single party rule to multiparty democracy.

DSP Division Spéciale Présidentielle. Mobutu's private army, this elite military unit was recruited almost entirely from the president's equatorial region. In stark contrast with the FAZ, its fighters were better paid and properly equipped.

FAZ Forces Armées Zairoises. The regular Zairean army. Rarely paid and barely trained, the FAZ's lack of discipline and cowardice were so notorious, Congolese citizens would pun that it was 'défazé' ('out of it').

Lingala The lingua franca of Congo, it is also the adjective used in Africa to refer to the country's music.

MIBA Minière du Bakwanga. State-controlled diamond mining operation based in the town of Mbuji Mayi.

MPR Mouvement Populaire de la Révolution. The party set up by Mobutu. Until the declaration of multiparty democracy, every Congolese was supposed to be a member.

RPF Rwandese Patriotic Front. The Tutsi-led rebel group that won control of Rwanda in the wake of the 1994 genocide masterminded by Hutu extremists.

SNIP Service Nationale d'Intelligence et de Protection, one of the many incarnations of the country's intelligence services. Under the stewardship of the Terminator, the sinister individuals who worked for it were known as 'the owls', a reference to their predilection for nocturnal visits.

UNITA União Nacional para a Independência Total de Angola. Angolan rebel movement led by Jonas Savimbi, dedicted to the overthrow of the former Marxist government in Luanda. Its leaders were on good terms with Mobutu, whose country acted as a conduit for US arms deliveries and a useful rear base for UNITA fighters trying to avoid disarming as required under a UN peace deal.

Bibliography

Further Reading

For a gripping, impeccably researched account of King Leopold's exploitation of the Congo, *King Leopold's Ghost* written by Adam Hochschild and published by Houghton Mifflin Company in 1998 is unbeatable. Hochschild focuses on the individuals who brought Leopold's barbarity to public awareness, often at considerable personal cost, including British journalist Edmund Morel, diplomat Roger Casement and black Americans George Washington Williams and William Sheppard.

The White Nile and *The Blue Nile* by Alan Moorehead, published by Penguin in 1962 and reissued many times since, sets Henry Morton Stanley's exploration of the Congo in the context of the West's gradual discovery of the African continent. Stanley is just one of the many driven explorers, curious aristocrats and obsessed missionaries who feature in an atmospheric, often highly moving account.

The River Congo by Peter Forbath, published by Houghton Mifflin Company in 1977, is the geographical and historical story of the great river. More narrowly focused in its subject matter than the Moorehead books, there are places where they overlap.

Stanley himself was a consummate journalist and knew how to tell a story with all the verve, style and dash required to reach the widest audience. *Through the Dark Continent*, Volumes One and Two, published in 1878, is a wonderful tale of an expedition into the unknown. *The Congo and the Founding of its Free State*, published in 1885, is a more eccentric and opinionated work, including a fascinating list of tips on how to survive the tropics. The *Autobiography* of Sir Henry Morton Stanley, published in 1909 and edited by his wife Dorothy Stanley, also gives a strong taste of the man.

Sean Kelly gives a readable and detailed exposition of the interventionist role the United States has played in Zaire in his *America's Tyrant*, published by the American University Press in 1993.

The Congo Cables by Madeleine Kalb, published by Macmillan in 1982, is a blow-by-blow account of the dramatic events before and after independence, from Lumumba's murder to Mobutu's takeover, as seen through the eyes of the Western ambassadors, UN officials and superpower leaders responding to one of the biggest crises of the Cold War. Currently out of print, it probably gives more detail than the ordinary reader requires.

The Rwanda Crisis—History of a Genocide by Gerard Prunier, published by Kampala's Fountain Press in 1995 and reissued since, remains the definitive account of Rwanda's 1994 genocide. Clear, authoritative and utterly compelling.

Background Material

In writing about the end of the regime I drew on material published in *Les Derniers Jours de Mobutu* (Éditions Gideppe) in 1998 by Honoré Ngbanda Ko Atumba, a fascinating account of Mobutu's final years by the former secret service chief and close aide; *La chute de Mobutu et l'effondrement de son armée* by exiled General Ilunga Shamanga (published privately) and *Dans la Cour de Mobutu* by son-in-law Pierre Janssen, published by Michel Lafon in 1997. *Les*

Dérives d'une Gestion Prédatrice by Professor Mabi Mulumba, the former premier, published in Kinshasa in 1998, was also helpful.

For those interested in the president himself, Mobutu attracted more than his fair share of hagiographers. In *Mobutu—Dignité pour l'Afrique*, published by Albin Michel in 1989, the president got the chance to tell his story to sympathetic journalist Jean-Louis Remilleux. Out of print now (published in the 1960s), but positively oozing admiration, are *Mobutu, L'Homme Seul* and *Mobutu: Le Point de Départ* by Francis Monheim, a Belgian journalist who covered the independence years. Leaning heavily in the opposite direction is *Le Dinosaure—le Zaire de Mobutu* by Colette Braeckman (Fayard, 1992), a Belgian journalist who has reported on events in central Africa for many years.

The problem with many of the books written about Zaire by Zaireans is that they are either turgid PhD theses unsuitable for general readers or are marred by personal score-settling. *Mobutu—l'Incarnation du Mal Zairois* by former prime minister and turncoat Nguz Karl i Bond, published by Rex Collings in 1982, is of historical interest. *Mobutu et l'Argent du Zaire*, written by former secret service man Emmanuel Dungia and published by l'Harmattan in 1993, is full of juicy tit-bits. Professor Isidore Ndaywel e Nziem is to be congratulated on his broad-ranging *Histoire Générale du Congo* (Duculot, 1998), a priceless reference work for anyone studying the country.

On the academic front, Crawford Young remains the authority in the English language, although you'll be hard put to find his lucid works on the shelves of contemporary bookshops. In French (and Flemish if you can read it), Jules Marchal is still battling to keep Belgium's past in the public eye. *L'État Libre du Congo—Volumes 1 and 2*, published in 1996 by Éditions Paula Bellings and *E. D. Morel Contre Leopold II* (L'Harmattan, 1996) are matter-of-fact and scrupulously researched accounts of the colonial era.

Index